The Millennium Project

2011
STATE OF THE FUTURE

JEROME C. GLENN, THEODORE J. GORDON,
AND ELIZABETH FLORESCU

RECOMMENDATIONS

Invaluable insights into the future for the United Nations, its Member States, and civil society.
Ban Ki-moon, Secretary-General, United Nations

Useful resources for teachers and students of global and futures education programs.
Jennifer M. Gidley, President, World Futures Studies Federation

A must read for any decisionmaker with a long-term vision.
Enrique Peña Nieto, Governor of the State of Mexico

An indispensable guide to being trend literate in an increasingly complex world.
Alison Sander, The Boston Consulting Group, USA

Best I have ever seen on Creative Industries Trends.
Klaus Haasis, CEO, MFG Baden-Württemberg mbH, Stuttgart, Germany

The value and role of The Millennium Project and the **State of Future** *for the education of future generations is priceless.*
Professor Shamsaddin Hajiyev Gummat, Rector of Azerbaijan State Economic University
Chairman of the Azerbaijan Parliament's Committee for Science & Education

An important "finger on the pulse" essential for decisionmaking at all levels; global cities cannot afford to ignore this work.
Sithole Mbanga, CEO, South African Cities Network

High quality knowledge on the global challenges for world business leaders and the public.
Arantza Laskurain, General Secretary, MONDRAGON Corporation, Spain
(world's largest corporation of worker-owned companies)

A fascinating read for anyone who wonders what the world will look like in 2030, and beyond.
Andres Oppenheimer, *Miami Herald* **syndicated columnist, co-winner Pulitzer Prize**

The SOF report continues, year after year, to be the best introduction—by far—to a broad range of major global issues and long-term remedies.
Global Foresight Books

ISBN: 978-0-9818941-5-7
© 2011 The Millennium Project
    4421 Garrison Street, NW
    Washington, D.C. 20016-4055 U.S.A.

Library of Congress Control Number: 98-646672

+1-202-686-5179 (F/P)
info@millennium-project.org

The *2011 State of the Future* is a publication of The Millennium Project, an international participatory think tank established in 1996

Cover: Mauricio Teran Pinell, Student, Universidad Privada Franz Tamayo, Santa Cruz, Bolivia.

Print Section--Table of Contents

The *2011 State of the Future* is composed of two parts: print and CD. This print section contains the executive summary of each of the studies conducted in 2010–11. The enclosed CD of about 8,000 pages contains the cumulative work of The Millennium Project since 1996 and details of the studies included in the print section.

The Table of Contents of the CD section appears on the next page

CD Section—Table of Contents
See preceding page for Table of Contents of the Print Section

The enclosed CD of about 8,000 pages contains the cumulative work of The Millennium Project since 1996 and details of the studies included in this print section.

MILLENNIUM PROJECT NODE CHAIRS

The Millennium Project interconnects global and local perspectives through its Nodes (groups of individuals and institutions)

Argentina
Miguel Angel Gutierrez
Latin American Center for Globalization & Prospective
Buenos Aires, Argentina

Australia
Anita Kelleher
Designer Futures
Inglewood, Australia

Chris Stewart
Greenfields Partners
Melbourne, Australia

Azerbaijan
Reyhan Huseynova
Azerbaijan Future Studies Society
Baku, Azerbaijan

Ali M. Abbasov
Minister of Comm. & IT
Baku, Azerbaijan

Bolivia
Veronica Agreda
Franz Tamayo University
La Paz & Santa Cruz, Bolivia

Brazil
Arnoldo José de Hoyos and Rosa Alegria
São Paulo Catholic University
São Paulo, Brazil

Brussels-Area
Philippe Destatte
The Destree Institute
Namur, Belgium

Canada
David Harries
Foresight Canada
Kingston, ON, Canada

Central Europe
Pavel Novacek, Ivan Klinec, Norbert Kolos
Charles University
Prague, Czech Republic; Bratislava, Slovak Republic, Warsaw, Poland

Chile
Héctor Casanueva
Vice President for Research and Development
Pedro de Valdivia University
Santiago de Chili, Chile

China
Zhouying Jin
Chinese Academy of Social Sciences
Beijing, China

Rusong Wang
Chinese Academy of Sciences
Beijing, China

Colombia
Francisco José Mojica
Universidad Externado de Colombia
Bogotá, Colombia

Dominican Republic
Yarima Sosa
Fundación Global Democracia & Desarrollo, FUNGLODE
Santo Domingo, Dominican Republic

Egypt
Kamal Zaki Mahmoud Shaeer
Egyptian-Arab Futures Research Association
Cairo, Egypt

Finland
Juha Kaskinen
Finland Futures Academy, Futures Research Centre
Turku, Finland

France
Saphia Richou
Prospective-Foresight Network
Paris, France

Germany
Cornelia Daheim
Z_punkt GmbH The Foresight Company
Cologne, Germany

Greece
Stavros Mantzanakis
Emetris, SA
Thessaloniki, Greece

Gulf Region
Ali Ameen
Office of the Prime Minister
Kuwait City, Kuwait

India
Mohan K. Tikku
Futurist / Journalist
New Delhi, India

Iran
Mohsen Bahrami
Amir Kabir University of Technology
Tehran, Iran

Israel
Yair Sharan and Aharon Hauptman
Interdisciplinary Center for Technological Analysis & Forecasting
Tel Aviv University
Tel Aviv, Israel

Italy
Enrico Todisco
Sapienza University of Rome
Rome, Italy

Antonio Pacinelli
University G. d'Annunzio
Pescara, Italy

Japan
Shinji Matsumoto
CSP Corporation
Tokyo, Japan

Kenya
Katindi Sivi Njonjo
Institute of Economic Affairs
Nairobi, Kenya

Malaysia
Theva Nithy
Universiti Sains Malaysia
Penang, Malaysia

Mexico
Concepción Olavarrieta
Nodo Mexicano. El Proyecto Del Milenio, A.C.
Mexico City, Mexico

New Zealand
Wendy McGuinness
Sustainable Future Institute
Wellington, New Zealand

Peru
Julio Paz
IPAE
Lima, Peru

Fernando Ortega
CONCYTEC
Lima, Peru

Russia
Nadezhda Gaponenko
Russian Institute for Economy, Policy & Law
Moscow, Russia

Silicon Valley
John J. Gottsman
Clarity Group
Palo Alto CA, USA

South Africa
Geci Karuri-Sebina
SA Cities Network
Johannesburg, South Africa

Southeast Europe
Blaz Golob
Centre for e-Governance Development for South East Europe
Ljubljana, Slovenia

South Korea
Youngsook Park
UN Future Forum
Seoul, Korea

Spain
Ibon Zugasti
PROSPEKTIKER, S.A.
Donostia-San Sebastian, Spain

Turkey
Alper Alsan
Siemens A.S., & All Futurists Association
Istanbul, Turkey

United Arab Emirates
Hind Almualla
Knowledge and Human Development Authority
Dubai, UAE

United Kingdom
Martin Rhisiart
Centre for Research in Futures & Innovation
Wales, Pontypridd, United Kingdom

Venezuela
José Cordeiro
Sociedad Mundial del Futuro Venezuela
Caracas, Venezuela

Arts/Media-Node
Kate McCallum
c3: Center for Conscious Creativity
Los Angeles, California

Joonmo Kwon
Fourthirtythree Inc.
Seoul, South Korea

Experimental Cyber-Node
Frank Catanzaro
Arcturus Research & Design Group
Maui, Hawaii

The Millennium Project was sponsored in its 2010–11 research program by:

- **Army Environmental Policy Institute, U.S. Army**
- **Azerbaijan State Economic University**
- **City of Gimcheon (via UN Future Forum, South Korea)**
- **The Diwan of His Highness the Amir of Kuwait**
- **Rockefeller Foundation**
- **Universiti Sains Malaysia**
- **UNESCO**

with in-kind support from:

- **CIM Engineering**
- **Smithsonian Institution**
- **UNESCO**
- **World Future Society**

This is the fifteenth report in an annual series intended to provide a context for global thinking and improved understanding of global issues, opportunities, challenges, and strategies.

The purposes of The Millennium Project are to assist in organizing futures research, improve thinking about the future, and make that thinking available through a variety of media for consideration in policymaking, advanced training, public education, and feedback, ideally in order to accumulate wisdom about potential futures.

The Project is designed to provide an independent global capacity that is interdisciplinary, transinstitutional, and multicultural for early alert and analysis of long-range issues, opportunities, challenges, and strategies. The Project is not intended to be a one-time study of the future but to provide an ongoing capacity as an intellectually, geographically, and institutionally dispersed think tank. Feedback on this work is welcome and will help shape the next *State of the Future*.

Previous *State of the Future* reports are available in Arabic, Azerbaijani, Chinese, English, French, Korean, Persian, Romanian, Russian, Czech, and Spanish. See www.millennium-project.org, "Books and Reports."

Readers of the *State of the Future* may also be interested in the *Futures Research Methodology Version 3.0*, which is the largest internationally peer-reviewed collection of methods to explore the future ever assembled in one resource. The CD has 39 chapters totaling approximately 1,300 pages. Also available is *Futures*, a Spanish-English extended dictionary of over 880 futures concepts and methods.

THE MILLENNIUM PROJECT PLANNING COMMITTEE

√ Rosa Alegria. PUC-SP São Paulo Catholic Univ., São Paulo, Brazil
Alper Alsan, Siemens A.S. and All Futurists Association of Turkey, Istanbul, Turkey
Ali Ameen, Office of the Prime Minister, Government of Kuwait, Kuwait
Mohsen Bahrami, Amir Kabir University of Technology, Tehran, Iran
Eleonora Barbieri Masini, Gregorian University, Rome, Italy
Peter Bishop, Futures Studies, University of Houston, Houston TX, USA
Héctor Casanueva, Universidad Pedro De Valdivia, Santiago de Chile, Chile
Frank Catanzaro, Arcturus Research & Design Group, Maui HI, USA
José Luis Cordeiro, Sociedad Mundial del Futuro Venezuela, Caracas, Venezuela
George Cowan, Founder, Santa Fe Institute, Santa Fe NM, USA
Cornelia Daheim, Z_punkt GmbH The Foresight Company, Essen, Germany
Francisco Dallmeier, Biodiversity, Smithsonian Institution, Washington DC, USA
Philippe Destatte, Director General, The Destree Institute, Namur, Wallonia, Belgium
√ Elizabeth Florescu, Director of Research, The Millennium Project, Calgary AB, Canada
Nadezhda Gaponenko, Russian Institute for Economy, Policy and Law, Moscow, Russia
√ Jerome C. Glenn, Executive Director, The Millennium Project, Washington DC, USA
Michel Godet, Conservatoire d'Arts et Métiers, Paris, France
Blaz Golob, Director, South-East Europe Centre for e-Governance, Ljubljana, Slovenia
Theodore J. Gordon, Senior Fellow, The Millennium Project, Old Lyme CT, USA
John J. Gottsman, President, Clarity Group, Atherton CA, USA
Miguel A. Gutierrez, Director, Latin American Center for Globalization and Prospective, Buenos Aires, Argentina
David Harries, The Canadian Defence Academy, Kingston, Ontario
√ Hazel Henderson, Futurist, Author, and Consultant, St. Augustine FL, USA
Arnoldo José de Hoyos Guevara, PUC-SP São Paulo Catholic Univ., São Paulo, Brazil
Reyhan Huseynova, President, Azerbaijan Future Studies Society, Baku, Azerbaijan
√ Zhouying Jin, Chinese Academy of Social Sciences, Beijing, China
Geci Karuri-Sebina, Executive Manager, Programmes, SA Cities Network, Johannesburg, South Africa
Juha Kaskinen, Director, Finland Futures Academy, Finland Futures Research Centre, Turku, Finland
Anita Kelleher, Designer Futures, Inglewood, Australia
Kamal Zaki Mahmoud Shaeer, Secretary-General, Egyptian-Arab Futures Research Association, Cairo, Egypt
Stavros Mantzanakis, Partner, Emetris Consulting, Thessaloniki, Greece
Shinji Matsumoto, President, CSP Corporation, Tokyo, Japan
Kate McCallum, President, C:3 Center for Conscious Creativity, Los Angeles CA, USA
Wendy McGuinness, Chief Executive, Sustainable Future Institute, Wellington, New Zealand
Francisco José Mojica, Director, Centro de Pensamiento Estratégico y Prospectiva, Bogotá, Colombia
Hind Al Mualla, Knowledge and Human Development Authority, Dubai, UAE
Theva Nithy, Universiti Sains Malaysia, Penang, Malaysia
Katindi Sivi Njonjo, Institute of Economic Affairs, Nairobi, Kenya
Pavel Novacek, Palacky University, Olomouc, and Charles University, Prague, Czech Republic
Concepción Olavarrieta, President, Nodo Mexicano. El Proyecto Del Milenio, A.C., Mexico City, Mexico
Fernando Ortega, CONCYTEC, Lima, Peru
Youngsook Park, President, UN Future Forum, Seoul, Republic of Korea
Charles Perrottet, Principal, The Futures Strategy Group, Glastonbury CT, USA
Cristina Puentes-Markides, Pan American Health Organization, Washington DC, USA
David Rejeski, Director, Foresight and Governance, Woodrow Wilson Center, Washington DC, USA
Saphia Richou, President, Prospective-Foresight Network, Paris, France
Stanley Rosen, President, Association for Strategic Planning, Los Angeles CA, USA
Paul Saffo, Director of Foresight at Discern Analytics, San Francisco CA, USA
Mihaly Simai, Director, World Institute of Economics, Budapest, Hungary
Yarima Sosa, Assistant to the President, Government of the Dominican Republic
Rusong Wang, Chinese Academy of Natural Sciences, Beijing, China
Paul Werbos, Program Director, National Science Foundation, Arlington VA, USA
Ibon Zugasti, PROSPEKTIKER, S.A., Donostia-San Sebastian, Spain

Sponsor Representatives
Ali Ameen, Office of the Prime Minister, Government of Kuwait, Kuwait
William Cosgrove, UNESCO World Water Scenarios project, France
John Fittipaldi, Army Environmental Policy Institute, USA
Reyhan Huseynova, President, Azerbaijan Future Studies Society, Baku, Azerbaijan
Claudia Juech, Rockefeller Foundation, USA
Theva Nithy, Universiti Sains Malaysia, Penang, Malaysia

ACKNOWLEDGMENTS

The Chairs and Co-chairs of the 40 Millennium Project Nodes, plus their members who help select participants, translate questionnaires, initiate projects, review text, and conduct interviews, were essential for the success of the research conducted for the annual *State of the Future* report in this and previous years.

Jerome Glenn, Theodore Gordon, and Elizabeth Florescu were partners in the research for this volume, with research and administrative assistance from Kawthar Nakayima, Hayato Kobayashi, and John Young. Special acknowledgment is given for Theodore Gordon's quantitative and conceptual leadership in the further development and assessments of the State of the Future Index in Chapter 2; for Jerome Glenn's leadership on the cumulative research on the 15 Global Challenges in Chapter 1, the executive summary, and the conclusions; and for Elizabeth Florescu's research and organization of environmental security issues for Chapter 6. Principal members of the environmental security scanning team for the monthly environmental security reports were Andrew Blencowe, Elizabeth Florescu, Jerome Glenn, Theodore Gordon, Odette Gregory, David Harries, Robert Jarrett, Hayato Kobayashi, Sterling Wright, and John Young.

Major contributions to the 15 Global Challenges in Chapter 1 were made by Elizabeth Florescu, Jerome Glenn, Theodore Gordon, Hayato Kobayashi, Tom Murphy, and John Young. Additional contributors and reviewers include: Amara Angelica, Guillermina Baena, Dennis Bushnell, Frank Catanzaro, Jose Cordeiro, Cornelia Daheim, Nicola Dahlin, Elisa Dijkhuis, Jim Disbrow, Greg Folkers, Leon Fuerth, Gilberto Gallopin, Nadezhda Gaponenko, Edgar Goell, Margo de Groot-Coenen, Miguel Gutierrez, Tanja Hichert, Omar Martínez Legorreta, Patricia Leidl, Tere Márquez, John McDonald, Lourdes Melgar, Juraj Mesik, Concepción Olavarrieta, Cornelius Patscha, Cristina Puentes-Markides, Rafael Serrano, Kamal Zaki Mahmoud Shaeer, Mahaly Simai, Pera Wells, Paul Werbos, and Raquel Zabala.

The Egypt 2020 study in Chapter 3 was led by Kamal Zaki Mahmoud Shaeer, together with the Cairo Node, the Egypt Arab Futures Research Association, and its Collaborative Partners. The future arts/media study in Chapter 4 was led by Kate McCallum, together with the Arts/Media Node and C3: The Center for Conscious Creativity. The Latin American Scenarios 2030 team was initiated and led by Jose Cordeiro with the help of all the Latin American Nodes of The Millennium Project and with the continuous collaboration of Jerome Glenn, Theodore Gordon, and Elizabeth Florescu. The leaders for Scenario 1 were Jose Cordeiro (Venezuela) with Fernando Ortega (Peru) and Yarima Sosa (Dominican Republic); for Scenario 2 the leaders were Miguel Angel Gutierrez with Luis Ragno and Javier Vitale (Argentina) and Hector Casanueva (Chile); for Scenario 3, Concepcion Olavarrieta (Mexico) with Francisco Jose Mojica (Colombia); and Scenario 4, Rosa Alegria and Arnoldo Jose de Hoyos (Brazil) with Ibon Zugasti (Spain) and Veronica Agreda (Bolivia). A special thanks to Javier Vitale (Argentina) for the invitations database, to Oöna Bilbao (Dominican Republic) for translations, to Jose Vicente Boix Canto (Spain) for the statistical tables and appendixes, and to Barry Hughes and Jose Solorzano for their help with the International Futures model.

A special thank you to Susan Jette for her continued work on the annotated scenario bibliography in the CD; to Peter Yim, President of CIM Engineering for hosting the Project's Web site and internal email lists; and to Frank Catanzaro for experimental collaborative software applications.

Linda Starke provided editing of the print section and John Young provided proofreading assistance for several sections in both the print and CD sections. Kawthar Nakayima, Elizabeth Florescu, and Hayato Kobayashi did the production and layout of both the print and CD sections of this publication, and Mauricio Teran Pinell from the Bolivian Node provided the cover artwork.

Interns who have contributed over the past year include: Joel Aftreth, Oöna Bilbao, Andrew Blencowe, Gunel Cahangirova, Karin Eklund, Alexandra Florescu, Omar Gonzalo Merida, Paul Bonwoo Gu, Farah Hagiyeva, Shahriyar Hasanverdiyev, Kasandra Housley, Sungcho Jeon, Sungjoong Kang, David Lenett, Sean Seokim Lee, Reese McArdle, Kamila Mustafayeva, Manuela Nicosia, Petra Novackova, Priscilla Nzabanita, Sam Park, Rambod Peykar, Lisa Tan, Jonathan Venezia, JiYun Yang, and Mohammad Zia.

FOREWORD

This year's *State of the Future* is another extraordinarily rich distillation of information for thought leaders, decisionmakers, and all those who care about the world and its future. Readers will learn how their interests fit into the global situation and how the global situation may affect them and their interests.

The purpose of futures research is to systematically explore, create, and test both possible and desirable futures in order to improve decisions. Decisionmaking is affected by globalization; hence, global futures research is needed to inform decisions made by individuals, groups, and institutions.

Just as the person on top of the mast on old sailing ships used to point out the rocks and safe channels to the captain below for the smooth running of the ship through uncharted waters, so too futurists with foresight systems for the world can point out problems and opportunities to leaders around the world. The Millennium Project is one such system.

Because the issues and solutions of our time are increasingly transnational, transinstitutional, and transdisciplinary, The Millennium Project was created as a global participatory think tank of futurists, scholars, scientists, business planners, and policymakers who work for international organizations, governments, corporations, NGOs, and universities.

Futures research has had an uncomfortable relationship with most academic research. As the latter advances, it tends to narrow its scope of study. In contrast, futures research tends to broaden its scope of study as it advances, taking into account a broader range of factors that affect future possibilities. It is not a science; the outcome of futures studies depends on the methods used and the skills of the practitioners. Its methods can be highly quantitative (such as the State of the Future Index in Chapter 2), a combination of quantitative and qualitative (such as the research that led to the Latin America 2030 Scenarios in Chapter 5), or primarily analytical (such the scanning that produced the emerging environmental security issues in Chapter 6) and conjectural and intuitive (such as the judgments on the future of arts, media, and entertainment in Chapter 4). It helps to provide a framework to better understand the present and to expand mental horizons (such as the Global Challenges described in Chapter 1).

The *2011 State of the Future* provides an additional eye on global change. This is the fifteenth *State of the Future* report. It contains the 15-year cumulative research and judgments of more than 3,000 thoughtful and creative people. Over 700 people participated in last year's studies. The institutional and geographic demographics of the participants can be found in Appendix A in the CD.

The annual *State of the Future* is an information utility that people can draw from and adapt to their unique needs. It provides a global strategic landscape that public and private policymakers use to improve their own strategic decisionmaking and global understanding. Business executives use the research as input to their planning. University professors, futurists, and other consultants find this information useful in teaching and research. Sections of previous reports have been used as university and high school texts.

The *2011 State of the Future* comes in two parts: this print edition of a series of distilled versions of the 2010–11 research and the enclosed CD with complete details of The Millennium Project's research this year and over the past 15 years.

2010–11 research and the enclosed CD with complete details of The Millennium Project's research this year and over the past 15 years.

The CD version of the report, which contains about 8,000 pages, is designed to serve as a reference document. For example, the print Chapter 1 on the 15 Global Challenges allocates two pages to each Challenge, while the CD devotes more than 1,400 pages to them. On the CD, each Challenge has a comprehensive overview, more detailed regional views, suggested indicators to assess progress or lack thereof on addressing the challenge, and a set of actions and views about those actions suggested from previous Global Lookout Panels. The statements in the longer versions in the CD do not represent a consensus because they are a distillation of a range of views from hundreds of participants, rather than an essay by a single author. We sought and welcomed a diversity of opinions. Hence, some of the issues raised and recommended actions seem contradictory. In addition, there does not appear to be a cause-and-effect relationship in some of the statements, and some sound like political clichés, but these are the views of the participants that may be useful to consider in the policy process. Nevertheless, it does present a more coherent overview of the global situation and prognosis than we have found elsewhere.

The CD can also be used to search for the particular items needed in customized work. For example, all the African sections on each of the 15 Challenges could be assembled into one paper by cutting and pasting (and possibly adding to the content by searching for results on Africa in other chapters), providing one report on Global Challenges and Issues for Africa.

The Millennium Project's diversity of opinions and global view is ensured by the Nodes—groups of individuals and organizations that interconnect global and local perspectives. They identify participants, conduct interviews, translate and distribute questionnaires, and conduct research and conferences. It is through their contributions that the world picture of this report and indeed all of The Millennium Project's work emerges.

Through its research, publications, conferences, and Nodes, The Millennium Project helps to nurture an international collaborative spirit of free inquiry and feedback for increasing collective intelligence to improve social, technical, and environmental viability for human development. Feedback on any sections of the book is most welcome at <jglenn@igc.org> and may help shape the next *State of the Future*.

Jerome C. Glenn	Theodore J. Gordon	Elizabeth Florescu
Director	Senior Fellow	Director of Research
The Millennium Project	The Millennium Project	The Millennium Project

Box 1

What Is New in This Year's Report

➤ Executive Summary of this year's research and update on the global situation.

➤ Both the short and the long versions of the 15 Global Challenges were updated.

➤ The 2011 State of the Future Index and national examples from Kuwait and Timor-Leste.

➤ Initial assessment of 34 new objectives and policy directions in Egypt.

➤ A set of four alternative scenarios for the future of Latin America by 2030.

➤ International assessment of 32 possible seeds of the future of arts, media, and entertainment.

➤ Distillation of more than 300 items related to environmental security identified over the past year and full text of items identified since 2002 in CD Chapter 9.1.

➤ The CD includes details and research that support the print version; it also includes the complete text of previous Millennium Project research:

- Detailed description of each of the 15 Global Challenges.

- Evolution and computation of the State of the Future Index.

- Global exploratory, normative, and very-long range scenarios, along with an introduction describing their development.

- Concept and concrete applications of Collective Intelligence systems.

- Science & Technology and Global Energy Scenarios and supporting studies.

- Assessment of governance-related issues and future strategy units in selected governments.

- Environmental security definitions, threats, related treaties; UN military doctrine on environmental issues; potential military environmental crimes and the International Criminal Court; changing environmental security military requirements in 2010–25.

- Two studies to create indexes and maps of the status of sustainable development.

- An international review of the concept of creating a "Partnership for Sustainable Development."

- Study of factors required for successful implementation of futures research in decisionmaking.

- An Annotated Scenarios Bibliography of over 850 scenario sets totaling over 2,150 scenarios.

Executive Summary

The world is getting richer, healthier, better educated, more peaceful, and better connected and people are living longer, yet half the world is potentially unstable. Food prices are rising, water tables are falling, corruption and organized crime are increasing, environmental viability for our life support is diminishing, debt and economic insecurity are increasing, climate change continues, and the gap between the rich and poor continues to widen dangerously.

There is no question that the world can be far better than it is—IF we make the right decisions. When you consider the many wrong decisions and good decisions not taken—day after day and year after year around the world—it is amazing that we are still making as much progress as we are. Hence, if we can improve our decisionmaking as individuals, groups, nations, and institutions, then the world could be surprisingly better than it is today.

Now that the Cold War seems truly cold, it is time to create a multifaceted compellingly positive view of the future toward which humanity can work. Regardless of the social divisions accentuated by the media, the awareness that we are one species, on one planet, and that it is wise to learn to live with each other is growing, as evidenced by the compassion and aid for Haiti, Pakistan, and Japan; the solidarity with democracy movements across the Arab world; the constant global communications that connect 30% of humanity via the Internet; and the growing awareness that global climate change is everyone's problem to solve.

Fifty years ago, people argued that poverty elimination was an idealistic fantasy and a waste of money; today people argue about the best ways to achieve that goal within 50 years. Twenty-five years ago, people thought that civilization would end in a nuclear World War III; today people think everyone should have access to the world's knowledge via the Internet, regardless of income or ideology.

The *2011 State of the Future* offers no guarantee of a rosy future. It documents potentials for many serious nightmares, but it also points to a range of solutions for each. If current trends in population growth, resource depletion, climate change, terrorism, organized crime, and disease continue and converge over the next 50–100 years, it is easy to imagine an unstable world with catastrophic results. If current trends in self-organization via future Internets, transnational cooperation, materials science, alternative energy, cognitive science, inter-religious dialogues, synthetic biology, and nanotechnology continue and converge over the next 50–100 years, it is easy to imagine a world that works for all.

The coming biological revolution may change civilization more profoundly than did the industrial or information revolutions. The world has not come to grips with the implications of writing genetic code to create new lifeforms. Thirteen years ago, the concept of being dependent on Google searches was unknown to the world; today we consider it quite normal. Thirteen years from today, the concept of being dependent on synthetic life forms for medicine, food, water, and energy could also be quite normal.

Computational biophysics can simulate the physical forces among atoms, making medical diagnostics and treatment more individually accurate. Computational biology can create computer matching programs to quickly reduce the number of possible cures for specific diseases, with millions of people donating their unused computer capacity to run the matching programs (grid computing). Computational media allows extraordinary pixel and voxel detail when zooming in and out of 3D images. Computational engineering brings together the world's available information and computer models to rapidly accelerate efficiencies in design. All these are changing the nature of science, medicine, and engineering, and their acceleration is attached to Moore's law; hence, computational everything will continue to accelerate the knowledge explosion. Tele-medicine, tele-education, and tele-everything will connect humanity, the built environment, and computational everything to address our global challenges.

The earthquakes, tsunamis, and nuclear disasters in Japan exposed the need for global, national, and local systems for resilience—the capacity to anticipate, respond to, and recover from disasters while identifying future technological and social innovations and opportunities. Related to resilience is the concept of collective intelligence—maybe the "next big thing" to help us make better decisions (see CD Chapter 6).

After 15 years of The Millennium Project's global futures research, it is increasingly clear that the world has the resources to address its challenges. What is not clear is whether the world will make good decisions fast enough and on the scale necessary to really address the global challenges. Hence, the world is in a race between implementing ever-increasing ways to improve the human condition and the seemingly ever-increasing complexity and scale of global problems.

So, how is the world doing in this race? What's the score so far? A review of the trends of the 28 variables used in The Millennium Project's global State of the Future Index provides a score card on humanity's performance in addressing the most important challenges; see Box 2 and Figures 1 and 2.

Box 2. The world score card

Where we are winning

1. Improved water source (percent of population with access)
2. Literacy rate, adult total (percent of people age 15 and above)
3. School enrollment, secondary (percent gross)
4. Poverty headcount ratio at $1.25 a day (PPP) (percent of population) (low- and mid-income countries)
5. Population growth (annual percent) (A drop is seen as good for some countries, bad for others)
6. GDP per capita (constant 2000 US$)
7. Physicians (per 1,000 people) (surrogate for health care workers)
8. Internet users (per 1,000 people)
9. Infant mortality (deaths per 1,000 live births)
10. Life expectancy at birth (years)
11. Women in parliaments (percent of all members)
12. GDP per unit of energy use (constant 2000 PPP $ per kg of oil equivalent)
13. Number of major armed conflicts (number of deaths >1,000)
14. Undernourishment (percent of population)
15. Prevalence of HIV (percent of population 15–49)
16. Countries having or thought to have plans for nuclear weapons (number)
17. Total debt service (percent of GNI) (low- and mid-income countries)
18. R&D expenditures (percent of national budget)

Where we are losing

19. Carbon dioxide emissions (kt)
20. Global surface temperature anomalies
21. People voting in elections (percent of population)
22. Levels of corruption (15 largest countries)
23. People killed or injured in terrorist attacks (number)
24. Number of refugees (per 100,000 total population)

Where there is uncertainty

25. Unemployment, total (percent of total labor force)
26. Non-fossil-fuel consumption (percent of total)
27. Population in countries that are free (percent of total global population)
28. Forestland (percent of all land area)

Some data in Figures 1–3 had to be adjusted for graphic illustration purposes; those adjustments are indicated in the respective labels in brackets.

Figure 1. Where we are winning

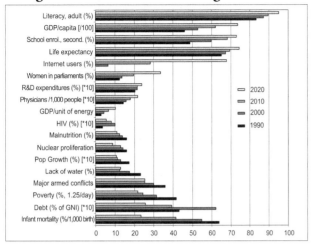

Figure 2. Where we are losing

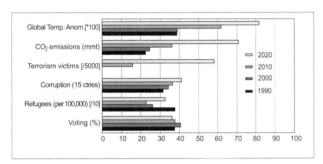

Figure 3. Where trends are not clear

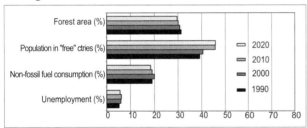

Figure 4. 2011 State of the Future

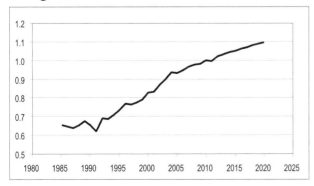

An international Delphi panel selected over a hundred indicators of progress or regress for the 15 Global Challenges in Chapter 1. Indicators were then chosen that had at least 20 years of reliable historical data and later, where possible, were matched with variables used in the International Futures model. The resulting 28 variables shown in Box 3 were integrated into the State of the Future Index with a 10-year projection. Chapter 2 in this print section presents a summary of this exercise, while full details are in Chapter 2 on the attached CD. SOFIs have also been computed for countries and could be applied to sectors like communications, health, water, and so forth.

The 2011 SOFI in Figure 4 shows that the 10-year future for the world is getting better. However, in many of the areas where we are winning we are not winning fast enough, such as reductions in HIV, malnutrition, and debt. And areas of uncertainty represent serious problems: unemployment, fossil fuel consumption, political freedom, and forest cover.

Some of the areas where we are losing could have quite serious impacts, such as corruption, climate change, and terrorism. Nevertheless, this selection of data indicated that 10 years from now, on balance, will be better than today.

Some Factors to Consider

Atmospheric CO_2 is at 394.35 ppm as of May 2011, the highest in at least 2 million years. Each decade since 1970 has been warmer than the preceding one; 2010 tied 2005 as the warmest year on record. The world is warming faster than the latest IPCC projections. Even the most recent estimates may understate reality, since they do not take into account permafrost melting.

According to FAO's *Livestock's Long Shadow* report, the meat industry adds 18% of human-related greenhouse gases, measured in CO_2 equivalent, which is higher than the transportation industry. A large reinsurance company found that 90% of 950 natural disasters in 2010 were weather-related and fit climate change models; these disasters killed 295,000 people and cost approximately $130 billion.

Humanity's material extraction increased eight times during the twentieth century. Today our consumption of renewable natural resources is 50% larger than nature's capacity to regenerate. In just 39 years, humanity may add an additional 2.3 billion people to world population. There were 1 billion humans in 1804; 2 billion in 1927; 6 billion in 1999; and 7 billion today. China is trying to become the green-growth giant of the world; it is too big to achieve reasonable standards of living for all its people first and then clean up later. Its next Five Year Plan (2011–15) allocated $600 billion for green growth initiatives.

Some believe the global ecosystem is crashing due to climate change, drying rivers and lakes, biodiversity loss, soil erosion, coastal dead zones, and collapsing bee populations unable to fertilize the food chain. Lester Brown in *Plan B 4.0* argues that nothing less than cutting CO_2 by 80% by 2020, keeping population to no more than 8 billion by 2050, restoring natural ecosystems, and eradicating poverty will save the ecosystem, and he proposes lowering income taxes as carbon taxes go up.

Since half of the largest 100 economies in the world are corporations, the former executive secretary of the UNFCCC argues that political leaders must give the business community a more central role in the transition to the green economy.

Falling water tables worldwide and increasing depletion of sustainably managed water have led some people to introduce the concept of "peak water," similar to peak oil. Fossil water – fossil fuels: both will peak, then what? It takes 2,400 liters of water to make a hamburger. Since 1990, an additional 1.3 billion people gained access to improved drinking water and 500 million got better sanitation. Yet 884 million people still lack access to clean water today (down from 900 million in 2009), and 2.6 billion people still lack access to safe sanitation. Half of all hospital patients in the developing world are there for water-related diseases.

As fertility rates fall and longevity increases, the ability to meet financial requirements for the elderly will diminish; the concept of retirement and social structures will have to change to avoid intergenerational conflicts. There were 12 persons working for every person 65 or older in 1950; by 2010, there were 9; and by 2050, the elderly support ratio is projected to drop to 4. There could be 150 million people with age-related dementia by 2050. Advances in brain research and applications to improve brain functioning and maintenance could lead to healthy long life, instead of an infirmed long life.

Food prices are the highest in history and are likely to continue a long-term trend of increases if there are no major innovations in production and changes in consumption, due to the combination of population growth, rising affluence (especially in India and China), the diversion of corn and other grains for biofuels, soil erosion, aquifer depletion, loss of cropland, falling water tables and water pollution, increasing fertilizer costs (high oil prices), market speculation, the diversion of water from rural to urban areas, increasing meat consumption, global food reserves at 25-year lows, and climate change's increasing droughts and flooding, melting mountain glaciers that reduce water flows, and eventually saltwater invading croplands. New approaches like saltwater agriculture, growing pure meat without growing animals, various forms of agro-ecology to reduce cost of inputs, and increasing vegetarianism would help.

Nearly 30% of the population in Moslem-majority countries is between 15 and 29 years old. Many who are without work and tired of older hierarchies, feeling left behind, and wanting to join the modern world brought change across North Africa and the Middle East this year. This demographic pattern is expected to continue for another generation, leading to both innovation and the potential for continued social unrest and migration.

The social media that helped the Arab Spring Awakening is part of a historic transition from many pockets of civilizations barely aware of each other's existence to a world totally connected via the current and future forms of the Internet. More data went through the Internet in 2010 than in all the previous years combined, and more electronic than paper books were sold by Amazon. Humanity, the built environment, and ubiquitous computing are becoming a continuum of consciousness and technology reflecting the full range of human behavior, from individual philanthropy to organized crime. New forms of civilization will emerge from this convergence of minds, information, and technology worldwide.

The number and percent in extreme poverty is falling. The world economy grew 4.9% in 2010 while the population grew 1.2%; hence, the world GDP per capita grew 3.7%. Nearly half a billion people rose out of extreme poverty ($1.25 a day) between 2005 and 2010. Currently this figure is about 900 million or 13% of the world. The World Bank forecasts this to fall to 883 million by 2015 (down from 1.37 billion in 2005). UNDP's new Multidimensional Poverty Index finds 1.75 billion people in poverty. In either case, the number of countries classified as low-income has fallen from 66 to 40. However, the gap between rich and poor within and among countries continues to widen. According to Forbes, the BRICs produced 108 of the 214 new billionaires in 2011. There are a total of 1,210 billionaires in the world now, of which 115 are citizens of China and 101 are Russian. The factors that increase the price of food, water, and energy are increasing; this has to be countered to address world poverty.

The world financial crisis and European sovereign debt emergencies continue to shift power to Asia, yet its leadership has not yet begun to help create that multifaceted general view of the future that humanity can work toward together. China became the second largest economy, passing Japan in 2010, and has more Internet users than the entire population of the United States. By 2030 India is expected to pass China as most populous country in the world. Together these two account for nearly 40% of humanity and are increasingly becoming the driving force for world economic growth.

World health is improving, the incidence of diseases is falling, and people are living longer, yet many old challenges remain and future threats are serious. During 2011 there were six potential epidemics. The most dangerous may be the NDM-1 enzyme that can make a variety of bacteria resistant to most drugs. New HIV infections declined 19% over the past decade; the median cost of antiretroviral medicine per person in low-income countries has dropped to $137 per year; and 45% of the estimated 9.7 million people in need of antiretroviral therapy received it by the end of 2010. Yet two new HIV infections occur for every person starting treatment. Over 30% fewer children under five died in 2010 than in 1990, and total mortality from infectious disease fell from 25% in 1998 to less than 16% in 2010. People are living longer, health care costs are increasing, and the shortage of health workers is growing, making tele-medicine and self-diagnosis via biochip sensors and online expert systems increasingly necessary.

Advances in synthetic biology, mail-order DNA, and future desktop molecular and pharmaceutical manufacturing could one day give single individuals the ability to make and deploy biological weapons of mass destruction. To counter this, advances in sensors to detect molecular changes in public spaces will be needed, along with advances in human development and social engagement to reduce the number of people who might be inclined to use these technologies for mass murder.

Another troubling area is the emerging problem of information and cyber warfare. Governments and military contractors are engaged in an intellectual arms race to defend themselves from cyberattacks from other governments and their surrogates. Because society's vital systems now depend on the Internet, cyberweapons to bring it down can be thought of as weapons of mass destruction. Information warfare's manipulation of media can lead to the increasing mistrust of all information.

Meanwhile, old style wars have decreased over the past two decades, cross-cultural dialogues are flourishing, and intra-state conflicts are increasingly being settled by international interventions. Today, there are 10 conflicts with at least 1,000 deaths per year (down from 14 last year): Afghanistan, Iraq, Somalia, Yemen, NW Pakistan, Naxalites in India, Mexican cartels, Sudan, Libya, and one classified as international extremism. The U.S. and Russia continue to reduce nuclear weapons while China, India, and Pakistan are increasing them. According to the Federation of American Scientists, by February 2011 there were 22,000 nuclear warheads, of which 2,000 are ready for use by the U.S. and Russia. The number and area of nuclear-free zones is increasing, but the number of unstable states grew from 28 to 37 between 2006 and 2011. Much of Central America could be called a failed or failing state in that organized crime controls people's lives more than governments do. Africa's population could double by 2050, with a growing number of unemployed youth and over 13 million AIDS orphans, increasing the likelihood of social instabilities and future conflicts.

With the potential collapse of Yemen, oil piracy along the Somali coast could increase. Ninety percent of international trade is carried by sea; 489 acts of piracy and armed robbery against ships were reported to IMO in 2010, up from 406 in 2009.

Investments into alternatives to fossil fuels are rapidly accelerating around the world to meet the projected 40–50% increase in demand by 2035. China has become the largest investor in "low-carbon energy," with a 2010 budget of $51 billion. Three Mile Island, Chernobyl, and now Japan's Fukushima nuclear disasters have left the future of that industry in doubt and strengthened the anti-nuclear movement in Japan and Europe.

Without major breakthroughs in technological and behavioral changes, the majority of the world's energy in 2050 will still come from fossil fuels. Therefore, large-scale carbon capture and reuse has to become a top priority to reduce climate change. Energy efficiencies, conservation, electric cars, tele-work, and reduced meat consumption are near-term ways to reduce energy GHG production. Automakers around the world are in a race to make lower-cost plug-in hybrid and all electric cars. Engineering companies are exploring how to take CO_2 emissions from coal power plants to make carbonates for cement and grow algae for biofuels and fish food. China is exploring tele-work programs to reduce long commuting, energy, costs, and traffic congestion.

Empowerment of women has been one of the strongest drivers of social evolution over the past century, and many argue that it is the most efficient strategy for addressing the global challenges in Chapter 1. Only two countries allowed women to vote at the beginning of the twentieth century; today there is virtually universal suffrage, the average ratio of women legislators worldwide has reached 19.2%, and over 20 countries have a woman head of state or government. Patriarchal structures are increasingly challenged, and the movement toward gender equality is irreversible.

Although the world is waking up to the enormity of the threat of transnational organized crime, the problem continues to grow, while a global strategy to address this global threat has not been adopted. World illicit trade is estimated at $1.6 trillion per year (up $500 billion from last year), with counterfeiting and intellectual property piracy accounting for $300 billion to $1 trillion, the global drug trade at $404 billion, trade in environmental goods at $63 billion, human trafficking and prostitution at $220 billion, smuggling at $94 billion, weapons trade at $12 billion, and cybercrime costing billions annually in lost revenue. These figures do not include extortion or organized crime's part of the $1 trillion in bribes that the World Bank estimates are paid annually or its part of the estimated $1.5–6.5 trillion in laundered money. Hence the total income could be $2–3 trillion—about twice as big as all the military budgets in the world.

The increasing complexity of everything in much of the world is forcing humans to rely more and more on computers. In 1997 IBM's Deep Blue beat the world chess champion. In 2011 IBM's Watson beat top TV quiz show knowledge champions. What's next? Just as the autonomic nervous system runs most biological decisionmaking, so too computer systems are increasingly making the day-to-day decisions for civilization.

The acceleration of S&T continues to fundamentally change the prospects for civilization, and access to its knowledge is becoming universal. Computing power and lowered costs predicted by Moore's Law continues with the world's first three-dimensional computer chip introduced by Intel for mass production. China currently holds the record for the fastest computer with Tianhe-1, which can perform 2.5 petaflops per second; IBM's Mira, ready next year, will be four times faster.

Is it possible that the acceleration of change will grow beyond conventional means of ethical evaluation? Will we have time to understand what is right and wrong as one change after the next makes it difficult to just keep up? For example, is it ethical to clone ourselves, or bring dinosaurs back to life, or invent new life forms from synthetic biology? These are not remote possibilities in a distant future; the knowledge needed to do them is being developed now. Despite the extraordinary achievements of S&T, future risks from their continued acceleration and globalization needs to be better forecasted and assessed. At the same time, new technologies also make it easier for more people to do more good at a faster pace than ever before. Single individuals initiate groups on the Internet, organizing actions worldwide around specific ethical issues. News media, blogs, mobile phone cameras, ethics commissions, and NGOs are increasingly exposing unethical decisions and corrupt practices, creating an embryonic global conscience. Our failure to inculcate ethics into more of the business community contributed to the global financial crisis and resulting recession, employment stagnation, and widening rich-poor gap.

Egypt 2020

The world cheered the Egyptian Revolution; now it wonders what's next. Will Egypt invent the first new form of democracy in the twenty-first century, taking into account the role of cyberspace, international interdependency, and a rapidly changing world? Will it become a centrally controlled political system with a decentralized local economic system? Or will it create a participatory democracy using the power of the Internet to constantly identify new approaches through a national collective intelligence system to address the persistent problems of poverty, water, education, and public health? It remains to be seen if the Arab Spring Awakening may eventually trigger a renaissance of Arabic and Islamic culture as they distinguish westernization from modernization. The Egyptian Node of The Millennium Project, together with the Egypt Arab Futures Research Association and its Collaborative Partners, created a Real-Time Delphi on the future of Egypt. Some highlights of the results are in Chapter 3, and the full study is available on the CD.

Future Arts, Media, and Entertainment

The explosive, accelerating growth of knowledge in a rapidly changing and increasingly interdependent world gives us so much to know about so many things that it seems impossible to keep up. At the same time, we are flooded with so much trivial news that serious attention to serious issues gets little interest, and too much time is wasted going through useless information. How can we learn what is important to know in order to make sure that there is a good future for civilization? Traditionally, the world has learned through education systems, art, media, and entertainment—and now with advances of communication and entertainment technologies, we have even more information and media at our fingertips on any number of ever-growing delivery systems.

Inspired by the Florentine Camerata Society, a sixteenth-century "think tank" responsible for the creation of the art form we know today as the European opera, The Millennium Project created the Arts and Media Node. The Node invited futuristic artists, media, and entertainment professionals and other innovators around the world to suggest and discuss future elements or seeds of the future of arts, media, and entertainment. After a month of online discussions, 34 elements were chosen and put into a Real-Time Delphi for an online international assessment. Writers, producers, performing artists, arts/media educators, and other professionals in entertainment, gaming, and communications were nominated by the 40 Millennium Project Nodes around the world to share their views. One distillation of the views of the participants shows that the future of arts, media, and entertainment will be a global, participatory, tele-present, holographic, augmented reality conducted on future versions of mobile smart phones that engage new audiences in the ways they prefer to be reached and involved. See Chapter 4 for a distillation of the results.

Latin America 2030

Between 2010 and 2030, most countries of Latin America will celebrate 200 years of independence in multiple bicentennial celebrations. Most countries in the region became independent following the French invasions of Portugal and Spain by Napoleon I in the early 1800s. As these countries look back over their first two centuries, it seemed appropriate to take this opportunity to explore future possibilities for Latin America. The Chairs of The Millennium Project Nodes in Latin America used a Real-Time Delphi that collected the judgments of 552 knowledgeable individuals about the likelihood and impacts of developments in Latin America over the next 20 years and the potential course of variables important to the region.

The results were used by four teams of Latin American Node Chairs to construct four scenarios: "Mañana" is Today: Latin American Success; Technology as Ideology: Believers and Skeptics; Region in Flames: This report is SECRET; and The Network: Death and Rebirth. Drafts of these four scenarios were shared via a Real-Time Delphi to collect feedback. The scenarios were then redrafted and are presented in Chapter 5. The full details of all the research that lead to the scenarios are available in the CD. All four scenarios are powerful resources for understanding the threats to and opportunities in the future of Latin America.

Environmental Security

Environmental security is increasingly dominating national and international agendas, shifting defense and geopolitical paradigms because it is increasingly understood that conflict and environmental degradation exacerbate each other. The traditional nation-centered security focus is expanding to a more global one due to geopolitical shifts, the effects of climate change, environmental and energy security, and growing global interdependencies.

The Millennium Project defines environmental security as environmental viability for life support, with three sub-elements: preventing or repairing military damage to the environment, preventing or responding to environmentally caused conflicts, and protecting the environment due to its inherent moral value.

Chapter 6 presents a summary of recent events and emerging environmental security–related issues organized around this definition. Over the past several years, with support from the U.S. Army Environmental Policy Institute, The Millennium Project has been scanning a variety of sources to produce monthly reports on emerging environmental issues with potential security or treaty implications.

More than 300 items have been identified during the past year and about 2,500 items since this work began in August 2002. The full text of the items and their sources, as well as other Millennium Project studies related to environmental security, are included in Chapter 9 on the CD and are available on The Millennium Project's Web site, www.millennium-project.org.

The *2011 State of the Future* ends with some brief conclusions. The readers are invited to draw their own conclusions and share them at mp-public@mp.cim3.net (after signing up at http://www.millennium-project.org/millennium/mp-public.html), The Millennium Project list on LinkedIn, or Twitter @MillenniumProj.

This year's *State of the Future* is an extraordinarily rich distillation of information for those who care about the world and its future. Since healthy democracies need relevant information, and since democracy is becoming more global, the public will need globally relevant information to sustain this trend. We hope the annual *State of the Future* reports can help provide such information.

The insights in this fifteenth year of The Millennium Project's work can help decisionmakers, opinion leaders, and educators who fight against hopeless despair, blind confidence, and ignorant indifference— attitudes that too often have blocked efforts to improve the prospects for humanity.

15 Global Challenges

The 15 Global Challenges provide a framework to assess the global and local prospects for humanity. The Challenges are interdependent: an improvement in one makes it easier to address others; deterioration in one makes it harder to address others. Arguing whether one is more important than another is like arguing that the human nervous system is more important than the respiratory system.

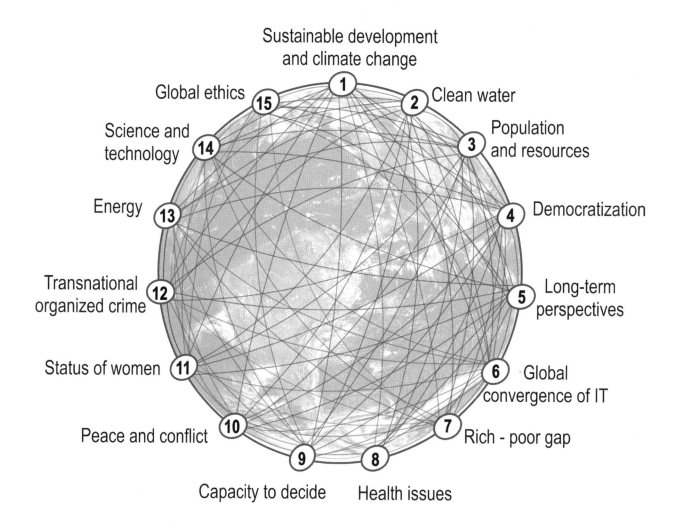

Readers are invited to contribute their insights to improve the overview of these 15 global challenges for next year's edition. Please use the online forms at www.StateoftheFuture.org (select "15 Global Challenges").

1

Global Challenges

This chapter presents two-page descriptions of 15 Global Challenges that have been identified and updated through an ongoing Delphi process and environmental scanning since 1996. The scanning process includes feedback from Millennium Project global assessments on the future of specific topics like media, energy, education, etc.; staff and interns scanning the Internet; expert reviews from the previous year's text; on-line feedback; regional input from The Millennium Project Node Chairs; feedback from The Millennium Project's email lists; monitoring conferences, seminars, and publications; and discussions around the world as staff and Node Chairs give talks on these challenges. All of this is distilled for patterns, and data are updated and cross-referenced.

These Challenges are transnational in nature and transinstitutional in solution. They cannot be addressed by any government or institution acting alone. They require collaborative action among governments, international organizations, corporations, universities, NGOs, and creative individuals. Although listed in sequence, Challenge 1 on sustainable development and climate change is no more or less important than Challenge 15 on global ethics. There is greater consensus about the global situation as expressed in these Challenges and the actions to address them than is evident in the news media.

More detailed treatments of the Global Challenges are available in the CD's Chapter 1, totaling over 1,400 pages. For each Challenge, there is a more comprehensive overview, alternative views or additional comments from participants on the overview, regional perspectives and relevant information from recent literature, a set of actions with a range of views from interviews with decisionmakers to address the challenge, additional actions and views on those actions, and suggested indicators to measure progress or lack thereof.

Both print and CD versions are the cumulative and distilled range of views from over 4,000 participants. See the Appendix for the demographics of the participants and see the CD's Appendix A for the full list of participants. Full details of the questionnaires and interview protocols that have been used from 1996 to 2009 to generate both the short and more detailed treatments of these Challenges are available at www.millennium-project.org (select "Lookout Studies").

The graphs at the end of this chapter illustrate trends for several variables and developments that assess changes relevant to the Global Challenges presented. They were created using the State of the Future Index methodology with trend impact analysis outlined in Chapter 2 and detailed in the CD Chapter 2.

1. How can sustainable development be achieved for all while addressing global climate change?

Atmospheric CO_2 is 394.35 ppm as of May 2011, the highest in at least 2 million years. Each decade since 1970 has been warmer than the preceding one; 2010 tied 2005 as the warmest year on record. The world is warming faster than the latest IPCC projections. Even the most recent estimates may understate reality since they do not take into account permafrost melting. Humans add about 45 gigatons of CO_2 equivalent of GHGs per year; half is processed by nature and half accumulates in the atmosphere. By 2050 another 2.3 billion people could be added to the planet and income per capita could more than double, dramatically increasing greenhouse gases. Climate change threatens the well-being of all humans, especially the poor, who have contributed the least to the problem. They are the most vulnerable to climate change's impacts because they depend on agriculture and fisheries, and they lack financial and technological resources to cope. The amount of global wealth exposed to natural disasters risk has nearly tripled from $525.7 billion 40 years ago to $1.58 trillion. Large reinsurance companies estimate the annual economic loss due to climate change could reach $300 billion per year within a decade.

Climate change could be accelerated by dangerous feedbacks: melting ice/snow on tundras reflect less light and absorb more heat, releasing more methane, which in turn increases global warming and melts more tundra; warming ocean water releases methane hydrates from the seabed to the air, warming the atmosphere and melting more ice, which further warms the water to release more methane hydrates; the use of methane hydrates or otherwise disturbing deeper sea beds releases more methane to the atmosphere and accelerates global warming; Antarctic melting reflects less light, absorbs more heat, and increases melting; and the Greenland ice sheet (with 20% of the world's ice) could eventually slide into the ocean.

The synergy between economic growth and technological innovation has been the most significant engine of change for the last 200 years, but unless we improve our economic, environmental, and social behaviors, the next 100 years could be disastrous. According to UNEP's *Towards a Green Economy* report, investing 2% of global GDP ($1.3 trillion per year) into 10 key sectors can kick-start a transition toward a low-carbon, resource-efficient green economy that would increase income per capita and reduce the ecological footprint by nearly 50% by 2050 compared with business as usual. Meanwhile, the world spends 1–2% of global GDP on subsidies that often lead to unsustainable resource use. The World Bank established a $100 million fund to support developing countries to set up their own carbon-trading scheme.

Glaciers are melting, polar ice caps are thinning, and coral reefs are dying. Some 30% of fish stocks have already collapsed, and 21% of mammal species and 70% of plants are under threat. Oceans absorb 30 million tons of CO_2 each day, increasing their acidity. The number of dead zones—areas with too little oxygen to support life—has doubled every decade since the 1960s. Over the long term, increased CO_2 in the atmosphere leads to proliferation of microbes that emit hydrogen sulfide—a very poisonous gas.

International negotiations on the post-Kyoto framework have shown insufficient progress since the voluntary national reduction targets of the Copenhagen Accord. UNEP estimates that these pledges would lead to a 20% overshoot in emissions in 2020 compared with the levels required to limit global warming to 2°C and stabilize at 450 ppm CO_2. There is also a growing fear that the target itself is inadequate—that the world needs to lower CO_2 to 350 ppm or else the momentum of climate change could grow beyond humanity's ability to reverse it. Emissions from increased production of internationally traded products have more than offset the emissions reductions achieved under the Kyoto Protocol.

Although the Montreal Protocol is expected to restore the ozone layer by 2050, depletion of that layer this spring reached an unprecedented level over the Arctic due to the continuing presence of ozone-depleting substances in the atmosphere and a very cold winter in the stratosphere.

Humanity's material extraction increased by eight times during the twentieth century. Today our consumption of renewable natural resources is 50% larger than nature's capacity to regenerate. Global ecosystem services are valued at $16–64 trillion, which far exceeds the sums spent to protect them.

It is time for a U.S.–China Apollo-like 10-year goal and global R&D strategy to address climate change, focusing on new technologies like electric cars, saltwater agriculture, carbon capture and reuse, solar power satellites (a Japanese national goal), pure meat without growing animals, maglev trains, urban systems ecology, and a global climate change collective intelligence to support better decisions and keep track of it all. These technologies would have to supplement other key policy measures, including carbon taxes, cap and trade schemes, reduced deforestation, industrial efficiencies, cogeneration, conservation, recycling, and a switch of government subsidies from fossil fuels to renewable energy.

Scientists are studying how to create sunshades in space, build towers to suck CO_2 from the air, sequester CO_2 underground, and reuse carbon at power plants to produce cement and grow algae for biofuels. Other suggestions include retrofitting coal plants to burn leaner and to capture and reuse carbon emissions, raising fuel efficiency standards, and increasing vegetarianism (the livestock sector emits more GHGs than transportation does). Others have suggested new taxes, such as on carbon, international financial transactions, urban congestion, international travel, and environmental footprints. Such taxes could support international public/private funding mechanisms for high-impact technologies. Massive public educational efforts via popular film, television, music, games, and contests should stress what we can do.

Given the difficulty of reaching a unanimous agreement, some argue that alternative forums such as G-20, the Montreal Protocol, or the Major Economies Forum may be a more realistic platform to manage climate change. Without a global strategy to address climate change, the environmental movement may turn on the fossil fuel industries. The legal foundations are being laid to sue for damages caused by GHGs.

Climate change adaptation and mitigation policies should be integrated into an overall sustainable development strategy. Without sustainable growth, billions more people will be condemned to poverty, and much of civilization could collapse, which is unnecessary since we know enough already to tackle climate change while increasing economic growth. Challenge 1 will be addressed seriously when green GDP increases while poverty and global greenhouse gas emissions decrease for five years in a row.

REGIONAL CONSIDERATIONS

AFRICA: The regional focus will be on adaptation to climate change rather than mitigation, as Africa does not contribute much CO_2. Southern Africa could lose more than 30% of its maize crop by 2030 due to climate change. Re-afforestation, saltwater agriculture along the coasts, and solar energy in the Sahara could be massive sources of sustainable growth.

ASIA AND OCEANIA: China is the world's largest CO_2 emitter, but it plans to cut the amount of energy and CO_2 per unit of economic growth by 16–17% from 2011 to 2016. Japan pledged to cut GHG emissions by 25% from 1990 levels by 2020, but its emissions are still well above the 1990 levels, and the government has failed to establish a domestic carbon trading market. More-stringent producer responsibility policies in South Korea triggered a 14% increase in recycling rates

and an economic benefit of $1.6 billion. China and India lose as much as 12% and 10% respectively of their GDP due to environmental damage. As part of a $1 billion deal with Norway, Indonesia introduced a two-year moratorium on new permits to clear primary forest.

EUROPE: Europe's emission trading scheme in 2010 accounted for 75% of the world's carbon emissions trading. GHG emissions covered under EU ETS increased 3% due to the economic recovery, but the EU is on track to meet the Kyoto target of 8% reduction. However, if carbon contents of imported goods are counted, EU's reduction drops to 1%.

Russia's GHG emissions fell 3.3% in 2009, reversing a 10-year steady increase, and Russia aims to reduce GHG emissions by 22–25% by 2020 compared with 1990 (which is still an increase in absolute terms, since Russia's emissions plunged sharply after the collapse of the Soviet Union). Nitrogen pollution from farms, vehicles, industry, and waste treatment costs the EU up to 320 billion euros per year. Germany tops the first Green Economy Index as a country with strongest green leadership.

LATIN AMERICA: South America has 40% of the planet's biodiversity and 25% of its forests. Brazil announced in December 2010 that deforestation in the Brazilian Amazon had fallen to its lowest rate for 22 years, but the latest data show a 27% jump in deforestation from August 2010 to April 2011, mostly in soybean areas. National pressures for hydro and biofuel energy and international pressures for food may be too strong to preserve ecosystems in Brazil. Concentration of land tenure, breakup of farms into smaller parcels, and conversion of rural areas into new urban settlements are generating irreversible ecological damage in most countries. Recycling in Brazil generates $2 billion a year, while avoiding 10 million tons of GHG emissions. Bolivia introduced a new law that grants nature equal rights to humans and has proposed an international treaty with similar concepts.

NORTH AMERICA: Without a successful green tech transition, U.S. GHG emissions may increase by 6% between 2005 and 2035. Air pollution and exposure to toxic chemicals cost U.S. children $76.6 billion in health expenses. Two-thirds of Latinos in the U.S. live in areas that do not comply with federal standards for air quality, and Hispanics are three times more likely than whites to die from asthma. Permafrost temperature in northern Alaska increased about 4–7°C during the last century, almost half of it during the last 20 years. On average, every American wastes 253 pounds of food every year. U.S. Congress refused to end oil subsidies.

2. How can everyone have sufficient clean water without conflict?

Water should be central to development and climate change strategies. Over half the world could live in water-stressed areas by 2050 due to population growth, climate change, and increasing demand for water per capita. According to IFPRI this would put at risk approximately $63 trillion of the global economy just 39 years from today. By 2030 global water demand could be 40% more than the current supply. This could change with new agricultural practices, policy changes, and intelligently applied new technologies. Otherwise conflicts over trade-offs among agricultural, urban, and ecological uses of water are likely to increase, along with the potential for mass migrations and wars. Although water-related conflicts are already taking place, water-sharing agreements have been reached even among people in conflict and have led to cooperation in other areas.

Today, some 2.4 billion people live in water-scarce regions. Falling water tables worldwide and increasing depletion of sustainably managed water lead some to introduce the concept of "peak water," similar to peak oil. Nevertheless, the world is on track to meet the MDG target on drinking water, but it is likely to miss the MDG sanitation target by almost 1 billion people. Since 1990, an additional 1.3 billion people gained access to improved drinking water and 500 million got better sanitation. Yet 884 million people still lack access to clean water today (down from 900 million last year), and 2.6 billion people still lack access to safe sanitation. Nature also needs sufficient water to be viable to support all life.

About 80% of diseases in the developing world are water-related; most are due to poor management of human excreta. At least 1.8 million children under five die every year due to unsafe water, inadequate sanitation, and the lack of hygiene. Diarrheal disease in children under 15 has a greater impact than HIV, malaria, and tuberculosis combined. WHO estimates that every dollar invested in improved sanitation and water produces economic benefits that range from $3 to $34, depending on the region and types of technologies applied.

Agriculture accounts for 70% of human usage of fresh water; the majority of that is used for livestock production. Such water demands will increase to feed growing populations with increasing incomes. Global demand for meat may increase by 50% by 2025 and double by 2050, further accelerating the demand for water per capita. The UN estimates that $50–60 billion annually between now and 2030 is needed to avoid future water shortages. Some 30% of global cereal production could be lost in current production regions due to water scarcity, yet new areas in Russia and Canada could open due to climate change. Cooling systems for energy production require large amounts of water. Energy demand may increase 40% in 20 years; coupled with increased food demands, dramatic changes in water management will be required.

Breakthroughs in desalination, such as pressurization of seawater to produce vapor jets, filtration via carbon nanotubes, and reverse osmosis, are needed along with less costly pollution treatment and better water catchments. Future demand for fresh water could be reduced by saltwater agriculture on coastlines, producing pure meat without growing animals, increasing vegetarianism, fixing leaking pipes, and the reuse of treated water.

Development planning should integrate the lessons learned from producing more food with less water via drip irrigation and precision agriculture, rainwater collection and irrigation, watershed management, selective introduction of water pricing, and successful community-scale projects around the world. Plans should also help convert degraded or abandoned farmlands to forest or grasslands; invest in household sanitation, reforestation, water storage, and treatment of industrial effluents in multipurpose water schemes; and construct eco-friendly dams, pipelines, and aqueducts to move water from areas of abundance to scarcity. Just as it has become popular to calculate someone's carbon footprint, people are beginning to calculate their "water footprint." The UN General Assembly declared access to clean water and sanitation to be a human right.

Challenge 2 will have been addressed seriously when the number of people without clean water and those suffering from water-borne diseases diminishes by half from their peaks and when the percentage of water used in agriculture drops for five years in a row.

REGIONAL CONSIDERATIONS

AFRICA: Africa's rapid urbanization has outpaced its capacity to provide sufficient water; the population without such access has nearly doubled since 1990 to over 55 million today. The ZAMCOM agreement to consolidate regional water management needs ratification by one more of the eight countries sharing the Zambezi river basin to come into effect. In the meantime, Mozambique, Zambia, and Zimbabwe moved ahead and signed a memorandum of understanding to improve power generation along the river. Foreign aid covers up to 90% of some sub-Saharan African countries' water and sanitation expenditures. Without policy changes, this region will not meet the MDG target on water until 2040 and the one on sanitation until 2076. Uganda launched a €212 Kampala Lake

Victoria Water and Sanitation project to upgrade and rehabilitate water supply and sanitation in urban and peri-urban Kampala. The "Safe Water for Africa" partnership plans to raise over $20 million to provide safe water to at least 2 million Africans by 2012. Since the majority of Africa depends on rain-fed agriculture, upgrading rain-fed systems and improving agricultural productivity will immediately improve millions of lives. Putting sanitation facilities in some village schools could bring girls back to school.

ASIA AND OCEANIA: Asia has 60% of the world's population but only 28% of its fresh water. Inadequate sanitation costs the economies of four Southeast Asian countries the equivalent of about 2% of their GDP. Agriculture accounts for between 65% and 90% of national water consumption across the Middle East, and underground aquifers are rapidly depleting. Yemen may have the first capital city to run out of water; it has the world's second-fastest growing population; its water tables are falling by 6.5 feet per year, and increasing water prices could spark social unrest. A massive infrastructure project plans to take water from the Yangtze River Basin and supply Beijing by 2014. In the meantime, the Beijing government will set aside $3.5 million to buy water from other provinces in 2011. More than 70% of China's waterways and 90% of its groundwater are contaminated; 33% of China's river and lake water is unfit for even industrial use; deep-groundwater tables have dropped by up to 90 meters in the Hai river basin. The water situation in China is expected to continue to get worse for the next six to nine years under the best-case scenarios. With only 8% of the world's fresh water, China has to meet the needs of 22% of the world's population. Forced migration due to water shortages has begun in China, and India should be next. The Yangtze, Mekong, Salween, Ganges, and Indus are among the 10 most polluted rivers in the world. India feeds 17% of the world's people on less than 5% of the world's water and 3% of its farmland. Mobile and nearly waterless public toilets that need to be cleaned only once a week will be piloted in Delhi, India. Saltwater intrusion into Bangladeshi coastal rivers reaches 100 miles inland and will increase with climate change. By 2050, an additional 1.5 billion m^3 of water will be needed in the Middle East, of which about a third will be allocated to the Palestinian Authority and Jordan. Chapter 10 of the Middle East Geneva Accords explains how to resolve Israeli-Palestinian water issues.

EUROPE: In 2009–10, water scarcity occurred in much of Southern Europe: the Czech Republic, Cyprus, and Malta reported continuous water scarcity; France, Hungary, the UK, Portugal, and Spain reported droughts or rainfall levels lower than the long-term average. The EU is to conduct a Policy Review for water scarcity and droughts in 2012. EU took Portugal to court for failing to submit river basin plans, an obligation under the EU Water Framework Directive. Water utilities in Germany pay farmers to switch to organic operations because it costs less than removing farm chemicals from water supplies. Spain is the first country to use the water footprint analysis in policymaking. The European Commission launched a €40-million fund to improve access to water in Africa, the Caribbean, and the Pacific. The world's largest reserves of fresh water are in Russia, which could export water to China and Middle Asia.

LATIN AMERICA: The region has 31% of the world's fresh water, yet 50 million people there have no access to safe drinking water, 125 million lack sanitation services, and 40% live in areas that hold only 10% of the region's water resources. The region's water demand could increase 300% by 2050. Mexico launched "2030 Water Agenda" for universal water access and wastewater treatment. Costa Rica needs to invest $2.4 billion to improve water and sanitation conditions by 2030. El Salvador will be hit hardest by water shortages in the region. Glaciers are shrinking, risking the region's water, agriculture, and energy security; 68% of the region's electricity is from hydroelectric sources. Water crises might occur in megacities within a generation unless new water supplies are generated, lessons from both successful and unsuccessful approaches to privatization are applied, and legislation is updated for more reliable, transparent, and consistent integrated water resources management.

NORTH AMERICA: The U.S. may have passed its "peak water" level in the 1970s. More than 30 states are in litigation with their neighbors over water. Some 13% of Native American households have no access to safe water and/or wastewater disposal, compared with 0.6% in non-native households. Each kilowatt-hour of electricity in the U.S. requires about 25 gallons of water for cooling, making power plants the second largest water consumer in the country (39% of all water withdrawals) after agriculture. Western Canada's tar sands consume an estimated 20–45 cubic meters of water per megawatt-hour, nearly 10 times that for conventional oil extraction. Canada is mapping its underground water supplies to help policymakers prevent water shortages. Government agricultural water subsidies should be changed to encourage conservation.

3. How can population growth and resources be brought into balance?

There were 1 billion humans in 1804; 2 billion in 1927; 6 billion in 1999; and 7 billion today. The UN's forecasts for 2050 range from 8.1 billion to 10.6 billion, with 9.3 billion as the mid-projection. Nearly all the population increases will be in urban areas in developing countries, where the slum population expanded from 767 million in 2000 to 828 million in 2010 and is expected to reach 889 million by 2020. Without sufficient nutrition, shelter, water, and sanitation produced by more intelligent human-nature symbioses, increased migrations, conflicts, and disease seem inevitable. ICT continues to improve the match between needs and resources worldwide in real time, and nanotech will help reduce material use per unit of output while increasing quality. However, food prices may continue to rise due to increasing affluence (especially in India and China), soil erosion and the loss of cropland, increasing fertilizer costs (high oil prices), market speculation, aquifer depletion, falling water tables and water pollution, diversion of crops to biofuels, increasing meat consumption, falling food reserves, diversion of water from rural to urban, and a variety of climate change impacts. The World Bank estimates that rising food prices pushed an additional 44 million people into poverty between June 2010 and January 2011.

Population dynamics are changing from high mortality and high fertility to low mortality and low fertility. If fertility rates continue to fall, world population could actually shrink to 6.2 billion by 2100, creating an elderly world difficult to support; if not, however, the UN projects 15.8 billion by 2100. Today life expectancy at birth is 68 years, which is projected to grow to 81 by 2100; with advances in longevity research, this projection will increase. About 20% of the world will be over 60 by 2050, and 20% of the older population will be aged 80 or more. Some 20% of Europeans are 60 or older, compared with 10% in Asia and Latin America and 5% in Africa. Over 20 countries have falling populations, which could increase to 44 countries by 2050, with the vast majority of them in Europe. Countering this "retirement problem" is the potential for future scientific and medical breakthroughs that could give people longer and more productive lives than most would believe possible today. People will work longer and create many forms of tele-work, part-time work, and job rotation to reduce the economic burden on younger generations and to keep up living standards.

Meanwhile, 925 million people were undernourished in 2010 (reduced from over 1 billion in 2009), while 30–40% of food production from farm to mouth is lost in many countries. The WFP provides food assistance to more than 90 million people in 73 countries, yet in some of these countries, agricultural lands (mostly in sub-Saharan Africa) are being sold or leased to foreign investors to feed their own countries. OECD estimates that the private sector's investment in farmland and agricultural infrastructure is as much as $25 billion and could double or triple over the next three to five years. Responsible Agricultural Investment, backed by the World Bank and UN agencies, aims to promote investment that respects local rights and livelihoods, but it is heavily criticized by NGOs as a move to legitimize land grabbing.

To keep up with population and economic growth, food production should increase by 70% by 2050. Meat consumption is predicted to increase from 37kg/person/year in 2000 to over 52kg/person/year by 2050; if so, then 50% of cereal production would go to animal feed.

Monocultures undermine biodiversity, which is critical for agricultural viability. Conventional farming relying on expensive inputs is not resilient to climatic change. Agricultural productivity could decline 9–21% in developing countries by 2050 as a result of global warming. Massive wheat damage by the Ug99 fungus in 2009 was less in 2010; its genome is now sequenced, leading to countermeasures, but it remains a threat according to FAO. Small-scale farmers can double food production within 10 years by using ecological methods. Agroecological farming projects have shown an average crop yield increase of 80% in 57 countries, with an average increase of 116% for all African projects. Aquaculture produces about half of human-consumed fish.

New agricultural approaches are needed, such as producing pure meat without growing animals, better rain-fed agriculture and irrigation management, genetic engineering for higher-yielding and drought-tolerant crops, precision agriculture and aquaculture, and saltwater agriculture on coastlines to produce food for human and animals, biofuels, and pulp for the paper industry as well as to absorb CO_2, reduce the drain on freshwater agriculture and land, and increase employment. An animal rights group has offered $1 million to the first producers of commercially viable in-vitro chicken by mid-2012.

Challenge 3 will be addressed seriously when the annual growth in world population drops to fewer than 30 million, the number of hungry people decreases by half, the infant mortality rate decreases by two-thirds between 2000 and 2015, and new approaches to aging become economically viable.

REGIONAL CONSIDERATIONS

AFRICA: FAO estimates that 20 million hectares of farmland have been acquired by foreign interests in Africa during the last three years, many with 50-year leases or more. About 40% of children under five are chronically malnourished. Very rapid growth of the young population and low prospects for employment in most nations in sub-Saharan Africa and some nations in the Muslim world could lead to prolonged instability until at least the 2030s. Africa's population doubled in the past 27 years to reach 1 billion and could reach 3.6 billion by 2100. Niger's population growth exceeds economic growth; if its birth rate is halved by 2050, the population will grow from 14 million today to 53 million by 2050, while if the birth rate continues at current levels the population will grow to 80 million. Much of the urban management class is being seriously reduced by AIDS, which is also lowering life expectancy. Only 28% of married women of childbearing age are using contraceptives, compared with the global average of 62%. Africa's ecological footprint could exceed its biocapacity within the next 20 years. Conflicts continue to prevent development investments, ruin fertile farmland, create refugees, compound food emergencies, and prevent better management of natural resources.

ASIA AND OCEANIA: Asia's urban population may grow to 3.1 billion by 2050. China has 88 cities with populations over 1 million. It plans to merge nine cities in the South to create a "mega-city" the size of Switzerland. China has to feed 22% of the world's population with less than 7% of the world's arable land. There were six Chinese children for every one elder in 1975; by 2035 there will be two elders for every one child. China is growing old before it has grown rich. The fertility rate in China has fallen from 5.8 children in 1970 to 1.5 today. By 2050, those 65 years or older will be 38% of Japan's population and 35% of South Korea's. Approximately a third of the population in the Middle East is below 15; another third is 15–29; youth unemployment there is over 25%. New concepts of employment may be needed to prevent political instability.

EUROPE: After 2012, the European working-age population will start to shrink, while the number of individuals aged 60 and over will continue to increase by about 2 million per year. Europe's low fertility rate and its aging and shrinking population will force changes in pension and social security systems, incentives for more children, and increases in immigrant labor, affecting international relations, culture, and the social fabric. The EU27 population at-risk-of-poverty has fluctuated around 16.5% since 2005. Tensions among the EU member states over the influx of thousands of illegal immigrants in the wake of "the Arab Spring" intensify as Mediterranean countries, led by Italy, ask for greater burden sharing and may lead to changes in the Schengen treaty. Rural populations are expected to shrink, freeing additional land for agriculture.

LATIN AMERICA: About 85% of the region will be urban by 2030, requiring massive urban and agricultural infrastructural investments. Over 53 million people are malnourished. Brazil, Ecuador, Venezuela, Guatemala, Honduras, and Nicaragua have approved food security laws to ensure local agricultural products are primarily used to feed their own populations and not for export; nine more countries are planning the same. Latin America's elderly population is likely to triple from 6.3% in 2005 to 18.5% in 2050—to 188 million. By 2050, half of Mexico's population will be older than 43, an 18-year increase in median age. As fertility rates fall in Brazil and longevity increases by 50% over the next 20 years, the ability to meet financial requirements for the elderly will diminish; hence, the concept of retirement will have to change and social inclusion will have to improve to avoid future intergenerational conflicts.

NORTH AMERICA: The number of those 65 or older in the U.S. is expected to grow from about 40 million in 2009 to 72 million in 2030. About 15% of American girls now begin puberty by age 7, potentially increasing girls' odds of depression and behavioral problems. Less than 2% of the U.S. population provides the largest share of world food exports, while 37 million people in the U.S. receive food from Feed America. Two-thirds of people in the U.S. are overweight or obese. Reducing "throw-away" consumption could change the population-resource balance. Biotech and nanotech are just beginning to have an impact on medicine; hence dramatic breakthroughs in longevity seem inevitable in 25–50 years. Vancouver, Toronto, and Calgary are among the five most livable cities of the world. Global warming should increase Canadian grain exports.

4. How can genuine democracy emerge from authoritarian regimes?

Peaceful protests with unrelenting public courage to demand democratic transitions from authoritarian regimes made history across the Arab world. Unparalleled forms of social power are shaping the future of democracy. Tensions between an expanding global consciousness and old structures that limit freedom are giving birth to new experiments in governance. Although the perception and implementation of democracy differ globally, it is generally accepted that democracy is a relationship between a responsible citizenry and a responsive government that encourages participation in the political process and guarantees basic rights.

Social revolutions in 2011 are not yet reflected in Freedom House's 2010 ratings, which showed political and civil liberties declined for the fifth consecutive year, the longest decline since 1972, when the annual analysis began. Freedom declined in 25 countries and improved in 11. Those living in 87 "free" countries constituted 43% of world population, while 20% live in 60 "partly free" countries, and 35% (over 2.5 billion people) live in 47 countries listed as "not free." There were 115 electoral democracies in 2010, compared with 123 in 2005. Press freedoms have declined for nine consecutive years; 15% of the world lives in the 68 countries with a "free" press, 42% in 65 countries with a "partly free" press, and 43% live in 63 countries without free media.

Predominantly young and increasingly educated populations are using the Internet to organize around common ideals, independent of conventional institutional controls and regardless of nationality or languages. These new forms of Internet-augmented democracy are beginning to wield unparalleled social power, often bypassing conventional news media, as happened in the Arab Spring Awakening, where 60% of the population is below the age of 30. And next? The transition to stable democracies will be difficult (see Chapter 3). New democracies must address previous abuses of power to earn citizens' loyalties without increasing social discord and slowing the reconciliation process.

Some global trends nurturing the emergence of democracy include increasing literacy, interdependence, Internet access, e-government systems, international standards and treaties, multipolarity and multilateralism in decisionmaking, developments that force global cooperation, improved quality of governance assessment systems, transparent judicial systems, and the growing number and power of NGOs. It is critical to establish legitimate tamper-proof election systems with internationally accepted standards for election observers. Some 20 countries offer legally binding Internet voting. Direct voting on issues via the Internet could be next to augment representative democracy.

Since an educated and informed public is critical to democracy, it is important to learn how to counter and prevent disinformation, cyberwarfare, politically motivated government censorship, reporters' self-censorship, and interest-group control over the Internet and other media. Organized crime, corruption, concentration of media ownership, corporate monopolies, increased lobbying, and impunity threaten democracy. Old ideological, political, ethnic, and nationalistic legacies also have to be addressed to maintain the long-range trend toward democracy. Fortunately, injustices in different parts of the world become the concern of others around the world, who then pressure governing systems to address the issue. Despite restrictions and intimidations, independent journalists, intellectuals, and concerned citizens are increasing global transparency via digital media.

Although making development assistance dependent on good governance has helped in some countries, genuine democracy will be achieved when local people—not external actors—demand government accountability. Since democracies tend not to fight each other and since humanitarian crises are far more likely under authoritarian than democratic regimes, expanding democracy should help build a peaceful and just future for all. Meanwhile, international procedures are needed to assist failed states or regions within states, and intervention strategies need to be designed for when a state constitutes a significant threat to its citizens or others.

Challenge 4 will be addressed seriously when strategies to address threats to democracy are in place, when less than 10% of the world lives in nondemocratic countries, when Internet and media freedom protection is internationally enforced, and when voter participation exceeds 60% in most democratic elections.

Regional Considerations

Africa: North African revolutions are not reflected in Freedom House's 2010 ratings. Ratings for sub-Saharan African democracy continued to decline; Ethiopia and Djibouti changed status to "not free," while only Guinea improved to "partly free." Freedom House rated 9 countries in the region as "free," 22 as "partly free", and 17 "not free." Democratic elections are still difficult due to intimidation and fraud. The Charter on Democracy, Elections and Governance adopted by the African Union in 2007 was signed by 37 AU Members; as of mid-May 2011 eight countries have ratified and another eight deposited the instruments of ratification, increasing the likelihood of increasing democratic values in the region. Priorities for building democracy in Africa include improving education, citizenry, and Internet access, while reducing corruption, sectarianism, violence, and patronage.

ASIA AND OCEANIA: Over the past few years, South Asia experienced more gains than setbacks, notes Freedom House. It rated 16 countries as "free" in the Asia-Pacific region, 15 as "partially free," and 8 as "not free." Notably successful elections took place in the Philippines and Tonga, while Sri Lanka suffered the most prominent decline in the region, due to its elections. Violent reprisal and censorship continue in several other countries. President Hu of China announced plans to improve its social management system by the end of the decade and turn China into a *xiaokang* (moderately prosperous and happy) society. Since the country is home to over half of the world population presently living in countries rated "not free," a modification of its status would change the world map of democracy. Among Central Asian countries, Kyrgyzstan's status improved from "not free" to "partly free," while Afghanistan continued to decline. In the Middle East and North Africa, Israel remains the only country ranked "free" and qualifying as an electoral democracy, while 3 countries are "partly free" and 14 "not free." However, the uprisings of 2011 open new possibilities for a more democratic society, despite the violent response of some countries' authoritarian regimes.

EUROPE: All 27 EU countries are rated "free," the EU Parliament is the largest transnational democratic electorate in the world, and the European Citizens' Initiative became law, enabling direct participation of citizens to propose regulations in areas under the Commission's authority. Yet, an increasing number of immigrants from Africa and Asia and their poor integration challenge the region's tradition of tolerance and civil liberties. Several member states call for a revision of the Schengen treaty on open borders. While the EU has the world's greatest press freedom, Hungarians protested against new media control legislation that some felt could return state censorship. In most Central and East

European (non-EU) countries, autocracy and lack of progressive institutions continues to hinder the democratization process. However, gains were noted in Georgia and Moldova, while Russia continues aggressive efforts to curb corruption.

LATIN AMERICA: The democratization of governments in the region is interrelated with the actions of the U.S. The big challenge for Latin America is the institutional weakness for addressing organized crime that is threatening its democracies. The interlinking of organized crime and government corruption caused Mexico's status to change in 2010 from "free" to "partly free." Freedom House rated 22 countries in the region "free," 10 "partly free," and 1 "not free." The system of primary elections in some countries favors those who are already in power and limits the freedom of choice for large majorities. However, a sense of solidarity of the people and increased influence of civil society organizations, as well as examples of democratic governance set by Chile and Brazil, are helping to strengthen democratic processes.

NORTH AMERICA: Concerns persist in Canada and the U.S. about the electoral processes, the concentration of media ownership, and powerful lobbies. Greater corporate and union spending on election advertising increases worries over political corruption. The U.S. State Department has budgeted $67 million to support democratic development in lower-income countries, while at home the future for 10–13 million illegal aliens challenges human rights and jurisprudence. Canada had four national elections in the past seven years. The Web site pairvote.ca is facilitating pairing voters from different voting districts to vote for each other's party, thus keeping the balance of popular vote unchanged while improving proportionate representation.

Figure 5. Global trends in freedom

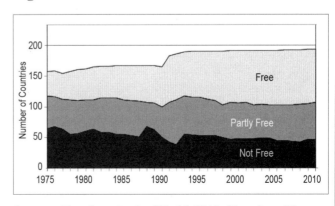

Source: Freedom in the World 2011, Freedom House

5. How can policymaking be made more sensitive to global long-term perspectives?

The earthquakes, tsunamis, and nuclear disasters in Japan exposed the need for global, national, and local systems for resilience—the capacity to anticipate, respond, and recover from disasters while identifying future technological and social innovations and opportunities. The Ministry of Foreign Affairs in Denmark notes that for every $1 invested in resilience and prevention, $4–7 are saved in response. Related to resilience is the concept of collective intelligence—emergent properties from synergies among brains, software, and information (see the CD Chapter 6), which will be increasingly required to cope with accelerating knowledge explosions, complexities, and interdependencies. Implementing and integrating resilience and collective intelligence systems is one way to make policymaking more sensitive to global long-term perspectives.

Heads of government could benefit from establishing an Office of the Future connected to related units in government agencies whose functions would continue from one administration to the next. These can be augmented by advisory councils of futurists and be connected to resilience and collective intelligence systems that scan for change around the world and can identify and assess expert judgments in real-time (the Real-Time Delphi is an example). The staff for such systems should synthesize futures research from others, calculate State of the Future Indexes for relevant subjects or countries (see Chapter 2), and produce annual state of the future reports. Existent government future strategy units (see the CD Chapter 4.1) are being networked by Singapore's Future Strategy Unit to share best practices, just as the UN Strategic Planning Network connects 12 UN agency strategy units. These two networks could also be connected with the Office of the UN Secretary-General to help coordinate strategies and goals. Leaders should make these new systems as transparent and participatory as possible to include and increase the public's intelligence and resilience. As a result, more future-oriented and global-minded voters might elect leaders who are sensitive to global long-term perspectives.

National legislatures could establish standing "Committees for the Future," as Finland has done. National foresight studies should be continually updated, improved, and conducted interactively with issue networks of policymakers and futurists and with other national long-range efforts. Futurists should create more useful communications to policymakers. Alternative scenarios should be shared with parliamentarians and the public for feedback. They should show cause-and-effect relations and expose decision points leading to different consequences from different strategies and policies. Decisionmakers and their advisors should be trained in futures research for optimal use of these systems (see www.millennium-project.org/millennium/FRM-V3.html). Government budgets should consider 5–10 year allocations attached to rolling 5–10 year SOFIs, scenarios, and strategies. Governments with short-term election cycles should consider longer, more-stable terms and funds for the staff of parliamentarians. A checklist of ways to better connect futures research to decisionmaking is available in Chapter 12 of the CD.

It could be that humanity needs and is ready to create a global, multifaceted, general long-range view to help it make better long-range decisions to the benefit of the species. Communications and advertising companies could create memes to help the public become sensitive to global long-term perspectives so that more future-oriented educated publics could support more future-oriented, global-minded politicians. Prizes could be given to recognize the best examples of global long-term decisionmaking. Participatory policymaking processes augmented by e-government services can be created that are informed by futures research. Universities should fund the convergence of disciplines, teach futures research and synthesis as well as analysis, and produce generalists in addition to specialists. Efforts to increase the number and quality of courses on futures concepts and methods should be supported, as well as augmenting standard curricula with futures methodologies converted to teaching techniques that help future-orient instruction.

Although there is increasing recognition that accelerating change requires global longer-term perspectives, decisionmakers feel little pressure to consider them. Nevertheless, attaining long-range goals like landing on the moon or eradicating smallpox that were considered impossible inspired many people to go beyond selfish, short-term interests to great achievements. (An international assessment of such future goals is found in Chapter 4.2 on the CD.) To some degree, the G20 was initiated to improve global long-range policymaking, and one day the G2 (U.S. and China) may lead global climate change and other long-range policies. Governments could add foresight as a performance evaluation criterion, add foresight to their training institutions, and require a "future considerations" section be added to policy reporting requirements. Promote the meme: from reaction to anticipation.

Each of the 15 Global Challenges in this chapter and the eight UN Millennium Development Goals could be the basis for transinstitutional coalitions composed of self-selected governments, corporations, NGOs, universities, and international organizations that are willing to commit the resources and talent to address a specific challenge or goal. Challenge 5 will be addressed seriously when foresight functions are a routine part of most organizations and governments, when national SOFIs are used in at least 50 countries, when the consequences of high-risk projects are routinely considered before they are initiated, and when standing Committees for the Future exist in at least 50 national legislatures.

REGIONAL CONSIDERATIONS

AFRICA: Foresightfordevelopment.org makes research documents, projects, scenarios, people, and blogs available to support African futures research. South Africa produces the regional Risk and Vulnerability Atlas to aid long-range planning. China has become a force in African long-range planning. Daily management of many African countries makes future global perspectives difficult; hence, more-regional bodies like the African Union and the African Development Bank are more likely to further futures work in Africa and should build on 10 years of work of UNDP/African Futures. Civil society is also becoming a bigger stakeholder and lobby in foresight.

ASIA AND OCEANIA: China's Five Year Plan promotes long-term thinking, and since it tends to make decisions in a longer time frame than others, its increasing power and eventually that of India should lead to more global, long-term decisionmaking as these nations interact with the rest of the world. Japan includes private-sector companies in its long-term strategic planning unit. The Prime Minister's Office of Singapore has begun an international network of government future strategy units.

EUROPE: EPTA, the European Parliamentary Technology Assessment, is a network and database of 18 European parliaments to integrate futures into decisionmaking. Forecasts of migrations from Asia and Africa are forcing Europe to reassess its future, as are the EU2020 strategy, Lisbon Strategy, sovereign debt crisis, emergence of China, and forecasts of public finances for social and health services for an aging population. The 7th Framework Programme of the EU expands foresight support; the Institute for Prospective Technological Studies provides futures studies for EU decisionmaking; the European Foresight Platform connects futurists; an annual European Futurists Conference is held in Switzerland; iKnow Project scans for weak signals and wild cards, and the European Regional Foresight College improves futures instruction. The Netherlands constitution requires a 50-year horizon for land use planning. Russian Ministries use Delphi and scenarios for foresight, while corporations tend to use technology roadmaps.

LATIN AMERICA: Research from ECLAC and UNIDO's technological foresight training could be improved to stimulate long-range decisionmaking; participation in such international organizations will improve the region's long-range global dialogs. Mexico initiated and signed an agreement to create the Pacific Latinamerican Alliance with the governments of Peru, Chile, and Colombia to promote free trade in a larger zone than Mercosur. Alternative long-term development strategies are being created by the Bolivian Alliance for the Americas, the Union of South American Nations, and the Community of Latin America and Caribbean States. The shift toward more socialist politics in some countries is motivating alternative futures thinking. Yet futures approaches are ignored by the academic and mass media, which focus on urgent and confrontational issues over ideologies, unmet basic needs, inequality, and large economic groups that monopolize services. Venezuela has the Sembrar el Futuro prize for students' futures thinking, and Mexico initiated the Global Millennium Prize for students' ideas for addressing global long-range challenges. Since the average age in Latin America is only 24, it is fundamental to incorporate the visions of the next generation via social networks and apps.

NORTH AMERICA: Create a map of individuals and organizations with foresight and use it to create a virtual organization at the White House for regular input to the policy process; the same for Langevin Block in Canada. "Future considerations" should be added to standard reporting requirements. Examples of successful global long-range activates should be promoted (see CD Chapter 12) along with cases where the lack of futures thinking proved costly. Global perspectives in decisionmaking are emerging due to perpetual collaboration among different institutions and nations that has become the norm to address the increasing complexity and speed of global change. Global long-term perspectives continue to be evident in the climate change policies of many local governments.

6. How can the global convergence of information and communications technologies work for everyone?

Over 2 billion Internet users, 5 billion mobile phones, and uncountable billions of hardware devices are intercommunicating in a vast real-time multi-network, supporting every facet of human activity. New forms of civilization will emerge from this convergence of minds, information, and technology worldwide. The eG8 was created in 2011 to explore government-business roles in managing this evolution. It is reasonable to assume that the majority of the world will experience ubiquitous computing and eventually spend most of its time in some form of technologically augmented reality. Today mobile devices have become personal electronic companions, combining computer, GPS, telephone, camera, projector, music player, TV, and intelligent guides to local and global resources.

As Moore's Law continues, costs fall, and ease of use increases, even remote and less developed areas will participate in this emerging globalization. The race is on to complete the global nervous system of civilization. Collaborative systems, social networks, and collective intelligences are self-organizing into new forms of transnational democracies that address issues and opportunities. This is giving birth to unprecedented international conscience and action, augmenting conventional management. Such open systems seem natural responses to increasing complexity that has grown beyond hierarchical control. Open source software's non-ownership model may become a significant element in the next economic system. Businesses are building offices and holding meetings in cyberworlds that compete with conventional reality.

One of the next "big things" could be the emergence of collective intelligences for issues, businesses, and countries, forming new kinds of organizations able to address problems and opportunities without conventional management. Collective intelligence can be thought of as a continually emerging property that we create (hands on) from synergies among people, software, and information that continually learns from feedback to produce just-in-time knowledge for better decisions than any one of these elements acting alone. Real-time streamed communications shorten the time it takes from situational awareness to decisions. Search engines and Wikipedia give instant access to "all" the world's stored knowledge. The Web is evolving from the present user-generated and participatory system (Web 2.0) into Web 3.0, a more intelligent partner that has knowledge about the meaning of the information it stores and has the ability to reason with that knowledge. Most mobile phones being sold today have computer capabilities, with thousands of apps and access to cloud computing.

However, this explosive growth of Internet traffic, mainly from video streaming, has created a stress on the Net's capacities, requiring new approaches to keep up with bandwidth demand, while the ubiquity of the Internet in society makes its reliability critically vital. People and businesses are trusting their data and software to "cloud computing" on distant Net-connected servers rather than their own computers, raising privacy and reliability questions. The Amazon cloud data center's outage and Sony PlayStation's release of personal data for millions of users are examples. Even though Wikipedia has become the world's encyclopedia, it struggles to counter disinformation campaigns fought through its pages. Governments are wrestling with how to control harmful content. A vigorous debate continues on net neutrality, the doctrine that technical and economic factors for Net users should not be affected by considerations of equipment, type of user, or communications content.

Humanity, the built environment, and ubiquitous computing are becoming a continuum of consciousness and technology reflecting the full range of human behavior, from individual philanthropy to organized crime. Low-cost computers are replacing high-cost weapons as an instrument of power in asymmetrical warfare. Cyberspace is also a new medium for disinformation among competing commercial interests, ideological adversaries, governments, and extremists, and is a battleground between cybercriminals and law enforcement. The full range of cybercrimes worldwide is estimated at $1 trillion annually. Fundamental rethinking will be required to ensure that people will be able to have reasonable faith in information. We have to learn how to counter future forms of information warfare that otherwise could lead to the distrust of all forms of information in cyberspace.

It is hard to imagine how the world can work for all without reliable tele-education, tele-medicine, and tele-everything. Internet bases with wireless transmission are being constructed in remote villages; cell phones with Internet access are being designed for educational and business access by the lowest-income groups; and innovative programs are being created to connect the poorest 2 billion people to the evolving nervous system of civilization. Two million children now have OLPC (One Laptop Per Child) XOs. Social networking spurs the growth of political consciousness and popular power, as in the "Arab Spring." E-government systems allow citizens to receive valuable information from their leaders, provide feedback to them, and carry out needed transactions without time-consuming and possibly corrupt human intermediaries. Telemedicine capabilities are uniting doctors and patients across continents. E-government systems exist to some degree for the majority of the world; the UN conducts comparative assessments of the e-government status of its member states.

Developing countries and foreign aid should have broadband access as national priorities, to make it easier to use the Internet to connect developing-country professionals overseas with the development processes back home, improve educational and business usage, and make e-government and other forms of development more available. Challenge 6 will have been addressed seriously when Internet access and basic tele-education are free and available universally and when basic tele-medicine is commonplace everywhere.

REGIONAL CONSIDERATIONS

AFRICA: According to worldinternetstats.com, Internet penetration in Africa is 10.9%, up 25% since last year. There are 506 million mobiles, for 50% penetration. The new Main One and West Africa fiber-optic cables are cutting cost and increasing speed. Kenya's Digital Villages Project integrates Internet access, business training, and microcredit. FAO's Africa Crop Calendar Web site provides information for 130 crops. Tele-education, tele-medicine, and e-government will become more important as African professionals die of AIDS in increasing numbers.

ASIA AND OCEANIA: Asia has the largest share of the world's Internet users (42%) but only 20% penetration. China has about 420 million Internet users with nearly 280 million Internet-connected mobile phones. Controversies over control of Internet access continue in China. Vietnam, India, Turkey, and Iran have tightened controls on Internet access and content. Phones are being smuggled into North Korea to post reports on conditions. The UN continues to rate South Korea the top e-ready country, but that nation is struggling with video game addiction. Some 300,000 people in Bangladesh are learning English from the BBC. India is establishing e-government stations in rural villages.

EUROPE: About 70% of EU-27 households had access to the Internet in 2010. Finland has made 1 MB/s broadband a legal right for all Finns. The EU's Safer Internet Programme is working in 26 European countries to counter child pornography, pedophilia, and digital bullying. The EU policy is that Internet access is a right, but it can be cut off for misuse. Estonians (inside and outside their country) cast their votes for the Estonian parliament by mobile phones in March 2011. Macedonia is providing computers to all in grades 1–3.

LATIN AMERICA: About 34% of the region has Internet access. The region's children with Internet access will rise from 1.5 million today to 30 million by 2015. Uruguay is the first country to provide all primary students with their own Internet-connected laptop, followed by Costa Rica. Fulfilling the promise of these tools will require more serious attention to training. The Internet was of great assistance in dealing with the Haiti earthquake. Fiber optic cable has been laid between Cuba and Venezuela.

NORTH AMERICA: Free to all on the Internet, Google and Wikipedia are making the phrase "I don't know" obsolete. Wikipedia is educating the world with 3.7 million articles in English and lesser amount in nine other languages. Silicon Valley continues as a world leader in innovative software due to company policies like Google's that gives its employees 20% free time to create anything they want. This "20-pecent Time" is credited with half of Google's new products. The United States is in ninth place in the world in access to high broadband connections. Broadband development in rural and underserved areas was undermined by the financial crisis, but it is still a U.S. national priority. The U.S. Computer Emergency Readiness Team reported an estimated 39% increase in cyber-attacks against government computer networks in 2010.

Figure 6: Internet users in the World by Geographic Region, 2011 (millions)

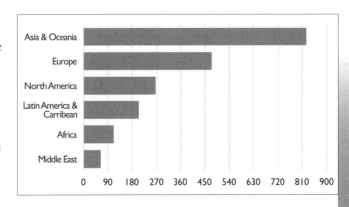

7. How can ethical market economies be encouraged to help reduce the gap between rich and poor?

Nearly half a billion people grew out of extreme poverty ($1.25 a day) between 2005 and 2010. The number and percent in extreme poverty is falling. Currently it is about 900 million or 13% of the world. The World Bank forecasts this to fall to 883 million by 2015 (down from 1.37 billion in 2005), while those living on less than $2 a day to fall to 2.04 billion from 2.56 billion. UNDP's new Multidimensional Poverty Index finds 1.75 billion people in poverty. The number of countries classified as low-income has fallen from 66 to 40. However, the gap between rich and poor within and among countries continues to widen due to globalization, some argue, and the number of unstable states grew from 28 to 37 between 2006 and 2011. Nevertheless, global economic recovery from the recession of 2009 (–0.5% world GDP) is expected to continue. The world's economy grew 5% in 2010, and the IMF expects growth at 4.5% for 2011 and 2012. Although better measures of human progress are being developed (see Chapter 2), GDP is still recognized as an indicator of economic change.

World GDP passed $74 trillion (PPP) in 2010 and is expected to pass $78 trillion in 2011, while per capita income grew from $10,800 (2009) to $11,100 (2010). The IMF forecasts economic growth to average 4.6% from 2011 to 2015, led mostly by emerging and developing economies. These are expected to average 6.6% growth compared with advanced economies averaging 2.5% growth over that same period. The contribution of BRIC to world GDP in 2010 was over 17.5%. And with the addition of South Africa to that group, this ratio is expected to increase to 33% by 2015. World trade shrank 12% in 2009 but grew 14.5% in 2010, and WTO expects it to grow another 6.5% in 2011. Developing economies' exports grew 16.5%. Remittances increased to $440 billion in 2010. The net ODA from DAC countries reached a record $128.7 billion in 2010, a 6.5% increase over 2009, but FDI inflows remained stagnant in 2010 at $1.1 trillion, and the ILO estimates that some 210 million people (about 32 million more than in 2007) are looking for jobs, bringing global unemployment to 6.2%. Industrial economies' unemployment accounted for 55% of the increase in worldwide unemployment between 2007 and 2010. UNFPA notes that in the 48 poorest countries, where population is expected to double by 2050, some 60% of the people are under age 25.

The world needs a long-term strategic plan for a global partnership between rich and poor. Forbes counts 1,210 billionaires, with 108 of the 214 added over the past year from the BRIC nations. Such a plan should use the strength of free markets and rules based on global ethics. Conventional approaches to poverty reduction (technical assistance and credit) that work in low- and middle-income stable countries do not work in fragile countries, which need stability first.

Ethical market economies require improved fair trade, increased economic freedom, a "level playing field" guaranteed by an honest judicial system with adherence to the rule of law and by governments that provide political stability, a chance to participate in local development decisions, reduced corruption, insured property rights, business incentives to comply with social and environmental goals, a healthy investment climate, and access to land, capital, and information. Direction from central government with relatively free markets is competing with the decentralized, individualized private enterprise for lifting people out of poverty.

New indicators for measuring progress and economic development are being developed to help managers move from short-term profit-based strategies to long-term viability. Technical assistance to leapfrog into new activities via tele-education and tele-work should be coupled with microcredit mechanisms for people to seek markets rather than non-existent jobs. An alternative to trying to beat the brain drain is to connect people overseas to the development process back home by a variety of Internet systems. If the WTO eliminated agricultural export subsidies, developing countries would gain $72 billion per year, according to UNDP. Structural imbalances in world trade have to be corrected to assure fair competition, respect of human rights, and labor and environmental standards, as well as efficient management of the global commons and prevention of monopolies. China's monetary policy adjustments could help other countries' economic development and access to world markets.

The low-carbon green economy has attracted over $2 trillion in private investment since 2007 and is a driving force for FDI. Climate Investment Funds of $6.4 billion help implement pilot projects in 45 developing countries through MDBs. Financing to the private sector by the MDBs increased from less than $4 billion in 1990 to $40 billion per year in 2010, while the IFC committed a record $18 billion in investments in private companies in 129 countries. Since 1976, microfinance institutions provided loans to over 113 million clients worldwide.

Challenge 7 will be addressed seriously when market economy abuses and corruption by companies and governments are intensively prosecuted and when the inequality gap—by all definitions—declines in 8 out of 10 years.

REGIONAL CONSIDERATIONS

AFRICA: Net bilateral ODA to Africa increased by 3.6% in 2010 to $29.3 billion, of which $26.5 billion went to sub-Saharan Africa. Despite a continued sustained economic growth since 2005 at an average of 4.7%, about half of sub-Saharan Africa continues to live in extreme poverty. Increasing commodity prices worldwide helped African oil exporters while having adverse effects on oil-importing countries. The East African Community is working toward economic integration. The rapidly evolving Chinese-African alliance is a new geopolitical reality that could help reduce income gaps for both sides; China-Africa trade is expected to triple between 2011 and 2015. The region's development continues to be impeded by high birth rates, increasing food prices, gender inequality, income and location biases, weak infrastructure, high indirect costs, corruption, armed conflicts, poor governance, environmental degradation and climate change, poor health conditions, and lack of education. Although the world will meet the MDG of halving poverty from 1990 to 2015 due to China's and India's growth, 17 African countries will not.

ASIA AND OCEANIA: Asia's economy grew 8.3% in 2010, according to the IMF, while WTO lists developing Asia's growth at 8.8%, and it is expected to grow over 8.4% per year over the next five years, led by China's average of 10% per year. India's poverty ($1.25/day), which was 51.3% in 1990, is expected to fall to 22.4% in 2015. China's poverty was 60% in 1990 and it is likely to plummet to 4.8% by 2015. China became the world's largest exporter (28% growth in 2010) and the second largest economy, with over 13% share of world economic output in 2010. China is now challenged to keep its growth from generating dangerous inflation. Japan's reconstruction after its environmental disasters will force it to reduce its development funding for the region. Increasing pollution, water and energy problems, and the rich-poor gap threaten the future economic growth of developing Asia. Corruption, organized crime, and conflict continue to impede Central Asia's development. Natural disasters and the effects of climate change are threatening the development and the very existence of entire Pacific communities.

EUROPE: EU-15 dynamic FDI to the other member states is fostering integration and helps economies across the EU. Despite signs of economic recovery, unemployment is expected to remain at around 10% for 2011–12; by March 2011, average youth unemployment in the Euro zone was 19.8% (but 44.6% in Spain). Cutbacks in social expenditures were protested across much of Europe and are likely to increase economic disparities. The European Financial Stabilization Mechanism to stabilize the euro and assist debt-stricken EU countries, along with the Europe 2020 Strategy, is intended to stimulate the regional economy; however, financial difficulties persist, causing friction during implementation. The combination of aging populations, falling fertility rates, a shrinking middle class in some countries, and expensive public services is not sustainable without increasing the number of immigrants and more tele-entrepreneurs among retired Europeans. In emerging Europe and Central Asia, 36% of the population lives on less than $5 per day. The Stabilization Fund helped Russia recover from the global financial crisis better than expected. It has one of the lowest foreign debts among major economies and its foreign reserves are the world's third largest, mainly due to revenues from oil and gas exports. Germany suggested lowering the EU's agricultural subsidies to improve foreign assistance.

LATIN AMERICA: The region's economy grew 6% in 2010, helped by rising commodity prices, 40% growth in 2010 capital flows (largest in history to the region), and stimulating policies. Regional GDP is expected to grow about 4% annually over 2011–15. FDI for Brazil increased by a record 87% in 2010. Yet the region's rich-poor gap continues as the world's largest. The wealthiest 20% manage 57% of resources, while the poorest 20% only get 3.5% of the income. Brazil, Mexico, and Argentina experience the highest inequalities. The region needs to attract high technology investments, create better access to the means of production, change land tenure, encourage international companies to increase salaries, create long-range visions for education and labor demand, and expand microcredit with business training.

NORTH AMERICA: The unemployment rate was 9% in the U.S. in April 2011 and 7.6% in Canada. The U.S. national debt is above the $14.3 trillion cap, and in 2010 over 43.9 million people (one in seven Americans) were enrolled in the food stamps program. Meanwhile, the top 0.1% of Americans control 10% of the nation's wealth, the U.S. has the most billionaires, and CEO pay rose 24%. The six largest U.S. banks control 63% of U.S. GDP, but new financial regulations give government more control over the banking system and financial markets and increase protection of the poor.

8. How can the threat of new and reemerging diseases and immune microorganisms be reduced?

World health is improving, the incidence of diseases is falling, and people are living longer, yet many old challenges remain and future threats are serious. Over 30% fewer children under five died in 2010 than in 1990, and total mortality from infectious disease fell from 25% in 1998 to less than 16% in 2010. Funding for global health continues to increase to an estimated $26.8 billion in 2010, which has also increased the need for better coherence among the many new actors in world health.

Non-communicable diseases and emerging and drug-resistant infectious diseases are increasing. Because the world is aging and increasingly sedentary, cardiovascular disease is now the leading cause of death in the developing as well as the industrial world. However, infectious diseases are the second largest killer and cause about 67% of all preventable deaths of children under five (pneumonia, diarrhea, malaria, and measles). Poverty, urbanization, travel, immigration, trade, increased encroachment on animal territories, and concentrated livestock production move infectious organisms to more people in less time than ever before and could trigger new pandemics. Over the past 40 years, 39 new infectious diseases have been discovered, 20 diseases are now drug-resistant, and old diseases have reappeared, such as cholera, yellow fever, plague, dengue fever, meningitis, hemorrhagic fever, and diphtheria. In the last five years, more than 1,100 epidemics have been verified. About 75% of emerging pathogens are zoonotic (they jump species).

During 2011 there were six potential epidemics. The most dangerous may be the NDM-1 enzyme that can make a variety of bacteria resistant to most drugs. Previously only found in hospitals, it was found this year in New Delhi's drinking water and sewers, making it easier to spread. Other notable developments this year include the European E. coli/ Hemolytic Uremic Syndrome foodborne outbreak; the progression of arteminisin-resistant malaria near the Cambodian border; the cholera epidemic in Haiti; the continued global threat from MDR and XDR TB in the AIDS population; and the potential for epidemics and nuclear contamination in post-earthquake, post-nuclear-meltdown Japan.

The H1N1 (swine flu) that infected millions around the world ended in August 2010 due to the ability of WHO and the global network to detect, isolate, genetically evaluate, vaccinate, and persuade the public to act. Mexico (where the virus was first identified) responded with praiseworthy professionalism in handling the A/H1N1 flu outbreak. The H5N1 (avian flu) of 2007–08 killed half of the people infected, spread slowly, has mutated three times in 15 years, and could mutate again. The best ways to address epidemic disease remain early detection, accurate reporting, prompt isolation, and transparent information and communications infrastructure, with increased investment in clean drinking water, sanitation, and handwashing. WHO's eHealth systems, international health regulations, immunization programs, and the Global Outbreak Alert and Response Network are other essentials of the needed infrastructure.

New HIV infections declined 19% over the past decade; AIDS-related deaths dropped by 19% between 2004 and 2009; the median cost of antiretroviral medicine per person in low-income countries has dropped to $137 per year; and 45% of the estimated 9.7 million people in need of antiretroviral therapy received it by the end of 2010. Yet two new HIV infections occur for every person starting treatment; 2.6 million were newly infected and 2 million died during 2009; and 33 million people are living with HIV/AIDS today. Some experts recommend a combination of annual voluntary universal testing in high-prevalence populations coupled with the immediate initiation of ART for those who test positive. Because ART reduces the viral load to the point where it cannot be detected it also prevents transmission. Others say this "test and treat" approach is too expensive and a human rights violation. WHO has adopted a new 2011–15 strategy instead that seeks to optimize HIV prevention, diagnosis, treatment, and care outcomes; to leverage broader health outcomes through HIV responses; to build strong and sustainable health systems; and to address inequalities and advance human rights.

Neglected tropical diseases are a group of parasitic and bacterial infections that are the most common afflictions of the world's poorest people. They blind, disable, disfigure, and stigmatize their victims, trapping them in a cycle of poverty and disease. Many low-cost interventions are available, yet the majority of affected people do not have access to them. Some of the largest health impacts remain: schistosomiasis (200 million cases), dengue fever (50 million new cases a year), measles (30 million cases a year), onchocerciasis (18 million cases in Africa), typhoid and leishmaniasis (approximately

12 million each globally), rotavirus (600,000 child deaths per year), and shigella childhood diarrhea (600,000 deaths per year). About half of the world's population is at risk of several endemic diseases. Hepatitis B infects up to 2 billion people. There is more TB in the world now than ever before, even though TB treatment success with DOTS exceeded 85%. Between 1995 and 2008, over 43 million people have been treated and 36 million people cured. There is progress with malaria: 38 countries (9 in Africa) documented reductions of more than 50% in the number of malaria cases between 2000 and 2008, and more than 100 million long-lasting insecticide-treated bed nets have been distributed in the fight against malaria.

To counter bioterrorism, R&D has increased for improved bio-sensors and general vaccines able to boost the immune system to contain any deadly infection. Such vaccines could be placed around the world like fire extinguishers. Some small viruses have been found to attack large viruses, offering the possibility of a new route to disease cures. New problems may come from unregulated synthetic biology laboratories of the future. People are living longer, health care costs are increasing, and the shortage of health workers is growing, making tele-medicine and self-diagnosis via biochip sensors and online expert systems increasingly necessary. Better trade security will be necessary to prevent increased food- or animal-borne disease. Viral incidence in animals is being mapped in Africa, China, and South Asia to divert epidemics before they reach humans. Future uses of genetic data, software, and nanotechnology will help detect and treat disease at the genetic or molecular level.

REGIONAL CONSIDERATIONS

AFRICA: With 11% of the world's population, Africa has 25% of the world's disease burden, 3% of its health workers, and 1% of its health expenditures. Sub-Saharan Africa accounted for 68% of all people living with HIV in 2010; it has one of the world's worst tuberculosis epidemics, compounded by rising drug resistance and HIV co-infection. Patients on ART increased from 1–2% in 2003 to 48% by the end of 2009. PEPFAR (a U.S. program) is funding 105 medical schools in the sub-Saharan region to encourage graduates to stay in Africa and is funding laboratories across the continent. Some 16% of children in Zimbabwe and 12% in Botswana are AIDS orphans; 34 sub-Saharan African countries stabilized or decreased HIV infections by more than 25% between 2001 and 2009.

ASIA AND OCEANIA: The emergent research on NDM-1 gene and drug resistance found in the New Delhi water system has alerted WHO investigators to a "potential nightmare" situation. Asia is an epicenter of emerging epidemics. If Asian poultry farmers received incentives to replace their live-market businesses—the source of many viruses—with frozen-products markets, the annual loss of life and economic impacts could be reduced. HIV continues to increase in central Asia. At least 5 million people have HIV/AIDS in India and China. Japan's life expectancy at birth in 2010 was 83 years; in China it was 74.

EUROPE: The Ukraine has the highest prevalence of HIV in Europe, focused on sex workers and drug users, with 161,119 cases (31,241 AIDS and 17,791 deaths), but it has decreased the incidence from 18% to 6% from 2006 to 2009 due to extensive HIV programs. The aging population of Europe continues to pressure government medical services, while infant mortality under five has been cut in half since 1990 and maternal mortality has dropped by one-fourth. TB deaths continue to increase in Europe after a 40-year decline. President Medvedev initiated obligatory drug tests in Russian schools and universities.

LATIN AMERICA: The region has the highest life expectancy among developing regions. While Haiti's HIV rate has fallen from 6% to 2.2% over the last 10 years, the earthquake killed 300,000 people and has devastated medical systems and brought on a cholera outbreak, with more than 1,200 deaths and the possibility of spread to the Americas. The HIV/AIDS epidemic remains stable, with 2 million people and 0.6% prevalence, and antiretroviral therapy is at almost 60%. Brazil has shown that free antiretroviral therapy since 1996 dramatically cut AIDS mortality, extended survival time, saved $2 billion in hospital costs, and keep prevalence to 0.6%. Neglected tropical diseases affect 200 million people in Latin America (intestinal worms, Chagas, schistosomiasis, trachoma, dengue fever, leishmaniasis, lymphatic filariasis, and onchocerciasis).

NORTH AMERICA: A California Biobank 20-year study will evaluate genetic markers for risk of disease in 250,000 patients by linking DNA samples to electronic medical records. The U.S. has 1.2 million people with HIV; Canada has 73,000. About 33% of children in the U.S. are overweight or obese, and one survey found that children aged 8–18 spent on average 7.5 hours a day with entertainment media.

9. How can the capacity to decide be improved as the nature of work and institutions change?

The increasing complexity of everything for much of the world is forcing humans to rely more and more on computers. In 1997 IBM's Deep Blue beat the world chess champion. In 2011 IBM's Watson beat top TV quiz show knowledge champions. What's next? Just as the autonomic nervous system runs most biological decisionmaking, so too computer systems are increasingly making the day-to-day decisions of civilization. We have far more data, evidence, and computer models to make decisions today, but that also means we have far more information overload and excessive choice proliferation. The number and complexity of choices seem to be growing beyond our abilities to analyze, synthesize, and make decisions. The acceleration of change reduces the time from recognition of the need to make a decision to completion of all the steps to make the right decision.

Many of the world's decisionmaking processes are inefficient, slow, and ill informed. Today's challenges cannot be addressed by governments, corporations, NGOs, universities, and intergovernmental bodies acting alone; hence, transinstitutional decisionmaking has to be developed, and common platforms have to be created for transinstitutional strategic decisionmaking and implementation. Previous economic models continue to mistakenly assume that human beings are well-informed, rational decisionmakers in spite of research to the contrary. And relying on computer models for decisions proved unreliable in the financial crisis. However, some progress has been made.

Adaptive learning models such as cellular automata, genetic algorithms, and neural networks are growing in capability and accuracy, and databases describing individual behavior are becoming even more massive. In social sciences it has been difficult to develop "laws" to forecast social behavior and, hence, make good decisions based on forecasted consequences. With the advent of massive digital databases and new software, we can let the computer make more empirically based forecasts of the plausible range of how people will react to various decisions. At the same time, increasing democratization and interactive media are involving more people in decisionmaking, which further increases complexity. This can reinforce the principle of subsidiarity—decisions made by the smallest number of people possible at the level closest to the impact of a decision. Fortunately, the world is moving toward ubiquitous computing with institutional and individual collective intelligence (emergent properties from synergies among brains, software, and information) for "just-in-time" knowledge to inform decisions. Ubiquitous computing will increase the number of decisions per day, constantly changing schedules and priorities. Decisionmaking will be increasingly augmented by the integration of sensors imbedded in products, in buildings, and in living bodies with a more intelligent Web and with institutional and personal collective intelligence software that helps us receive and respond to feedback for improving decisions.

Cloud computing, knowledge visualization, and a variety of decision support software are increasingly available at falling prices. DSS improves decisions by filtering out bias and providing a more objective assessment of facts and potential options. Some software lets groups select criteria and rate options, some averages people's bets on future events, while others show how issues have alternative positions and how each is supported or refuted by research.

The MIT Collective Intelligence Center sees its mission as answering "How can people and computers be connected so that collectively they act more intelligently than any individuals, groups, or computers have ever done before?" They are trying to develop measures of collective intelligence (like IQ tests for individuals). Rapid collection and assessment of many judgments via on-line software can support timelier decisionmaking. (See the attached CD Appendix L for an explanation of the Real-Time Delphi.) Google invited "citizen cartographers" to refine the U.S. map. This sort of activity is fundamentally different than the "wisdom of crowds" in which the average judgment is taken to be an answer to unresolved issues. The "wisdom of crowds" approach is essentially a vote, while collective intelligence is a continually emergent property from synergies among data-information-knowledge, software-hardware, and individual and groups of brains that continually learn from feedback. Self-organization of volunteers around the world via Web sites is increasing transparency and creating new forms of decisionmaking. Blogs are increasingly used to support decisions. Issues-based information software in e-government allows decisionmaking to be more transparent and accountable. Although cognitive neuroscience promises to improve decisionmaking, little has been applied for the public.

Political and business decisions include competitive intelligence and analysis to guide decisionmaking; as the world continues to globalize, increasing interdependencies, synergetic intelligence and analysis should also be considered. What synergies are possible among competing businesses, groups, and nations? Synergetic analysis aims to increase "win-win" decisions that assist a larger number of enterprises while reducing the wasted efforts of "win-lose" decisions.

Often decisions are delayed because people don't know something—a condition Google is beginning to eliminate. Training programs for decisionmakers should bring together research on why irrational decisions are made, lessons of history, futures research methods, forecasting, cognitive science, data reliability, utilization of statistics, conventional decision support methods (e.g., PERT, cost/benefit, etc.), collective intelligence, ethical considerations, goal seeking, risk, the role of leadership, transparency, accountability, participatory decisionmaking with new decision support software, e-government, ways to identify and better an organization's improvement system, prioritization processes, and collaborative decisionmaking with different institutions.

Challenge 9 will be addressed seriously when the State of the Future Index or similar systems are used regularly in decisionmaking, when national corporate law is modified to recognize transinstitutional organizations, and when at least 50 countries require elected officials to be trained in decisionmaking.

REGIONAL CONSIDERATIONS

AFRICA: North African revolutions promise to open the decisionmaking processes, increasing freedom of the press to better inform the public. For tribally oriented Africa, the question remains, how can the cultural advantages of extended families be kept while making political and economic decisions more objective and less corrupt? Development of African civil society may need external pressure for freedom of the press, accountability, and transparency of government. Microsoft is collaborating to help e-government systems improve transparency and decisionmaking. If the brain drain cannot be reversed, expatriates should be connected to the development processes back home through Internet systems.

ASIA AND OCEANIA: In general, decisions tend to focus more on the good of the family than on the good of the individual in Asian societies; will individualistic Internet change this philosophy? Synergies of Asian spirituality and collectivist culture with more linear, continuous, and individualistic western decisionmaking systems could produce new decisionmaking philosophies. Kuwait is introducing

a collective intelligence system and a national SOFI for the Early Warning System in the Prime Minister's Office. ASEAN could be the key institution to help improve decisionmaking systems in the region.

EUROPE: Bureaucratic complexity, lack of transparency, and proliferation of decision heads threatens clear decisionmaking in the EU. Europe is experiencing "reporting fatigue" due to so many treaties and bureaucratic rules. Tensions between the EU and its member governments and among ethnic groups are making decisionmaking difficult. Russia is improving policy decisionmaking efficiency by coordination among stakeholders in nanotechnology research among several Councils, Commissions at the Russian Parliament, government, and the Russian Academy of Science. It was a response to the cross-sectoral and multidisciplinary nature of nanotech.

LATIN AMERICA: Chile is pioneering e-government systems that can be models for other countries in the region. For e-government to increase transparency, reduce corruption, and improve decisions, Internet access beyond the wealthiest 20% is necessary. The remaining 80% receive inefficient service, difficult access locations, restricted operating hours, and non-transparent processes. Government institutional design, management, and data for decisionmaking are weak in the region. Latin America has to improve citizen participation and public education for political awareness.

NORTH AMERICA: Blogs and self-organizing groups on the Internet are becoming de facto decisionmakers in North America, with decisions made at the lowest level appropriate to the problem. Approximately 20% of U.S. corporations use decision support systems to select criteria, rate options, or show how issues have alternative business positions and how each is supported or refuted by research. Intellipedia provides open source intelligence to improve decisionmaking. The region's dependence on computer-augmented decisionmaking—from e-government to tele-business—creates new vulnerabilities to manipulation by organized crime, corruption, and cyber-terrorism, as discussed in Challenges 6 and 12.

10. How can shared values and new security strategies reduce ethnic conflicts, terrorism, and the use of weapons of mass destruction?

Although the vast majority of the world is living in peace, half the world continues to be vulnerable to social instability and violence due to growing global and local inequalities, outdated social structures, inadequate legal systems and increasing costs of food, water, and energy. In areas of worsening political, environmental, and economic conditions, increasing migrations can be expected. Add in the future effects of climate change, and there could be as many as 400 million migrants by 2050. While inter-state conflicts decreased, internal unrest is increasing. The UN estimates that 40% of the internal conflicts over the past 60 years were natural resource–related. As growing populations and economies increase the drain on natural resources, social tensions are expected to increase, triggering complex interactions of old ethnic and religious conflicts, civil unrest, terrorism, and crime. Substantial technological and social changes will be needed to prevent this; countries will need to include non-traditional security strategies for addressing the root causes of unrest. Since many countries affected by conflict return to war within five years of a cease-fire, more serious efforts are required to dismantle the structures of violence and establish structures of peace.

Conflicts have decreased over the past two decades, cross-cultural dialogues are flourishing, and intra-state conflicts are increasingly being settled by international interventions. Today, there are 10 conflicts (down from 14 last year) with at least 1,000 deaths per year: Afghanistan, Iraq, Somalia, Yemen, NW Pakistan, Naxalites in India, Mexican cartels, Sudan, Libya, and one classified as international extremism. Yet the 27.5 million internally displaced persons is the highest total since the 1990s. The probability of a more peaceful world is increasing due to the growth of democracy, international trade, global news media, the Internet and new forms of social networks, NGOs, satellite surveillance, better access to resources, and the evolution of the UN and regional organizations. The U.S. and Russia signed the new START nuclear arms reduction treaty, and new arms races are being preemptively addressed. Yet the Global Peace Index's rating of 144 countries' peacefulness again declined slightly, reflecting intensification of some conflicts and the economic crisis.

In 2011, there are 122,000 UN peacekeepers from 114 countries in 15 operations. Total military expenditures are about $1.5 trillion per year. There are an estimated 8,100 active nuclear weapons, down from 20,000 in 2002 and 65,000 in 1985. However, there are approximately 1,700 tons of highly enriched uranium and 500 tons of separated plutonium that could produce nuclear weapons. The nexus of transnational extremist violence is changing from complex organized plots to attacks by single individuals or small independent groups. Mail-order DNA and future desktop molecular and pharmaceutical manufacturing, plus access (possibly via organized crime) to nuclear materials, could one day give single individuals the ability to make and use weapons of mass destruction—from biological weapons to low-level nuclear ("dirty") bombs. The IAEA reports that between 1993 and the end of 2010 the Illicit Trafficking Database confirmed 1,980 incidents of illicit trafficking and other unauthorized activities involving nuclear and other radioactive materials. During 2010, the IAEA received reports of 176 nuclear trafficking incidents (compared with 222 during 2009), ranging from illegal possession and attempted sale and smuggling to unauthorized disposal of materials and discoveries of lost radiological sources.

Governments and military contractors are engaged in an intellectual arms race to defend themselves from cyberattacks from other governments and their surrogates. Beyond defense, the rules of engagement for responding to such aggressors are not clear. Because society's vital systems now depend on the Internet, cyberweapons to bring it down can be thought of as weapons of mass destruction. The security requirements for deterring single individuals from making and deploying WMD is unprecedented. We have to develop mental health and education systems to detect and treat individuals who might otherwise grow up to use such advanced weapons, as well as using networks of nanotech sensors to alert authorities to those creating such weapons.

Military power has yet to prove effective in asymmetrical warfare without genuine cultural engagement. Peace strategies without love, compassion, or spiritual outlooks are less likely to work, because intellectual or rational systems alone are not likely to overcome the emotional divisions that prevent peace. Conflict prevention efforts should work in and with all the related factions, including conversations with hardliner groups, taking into consideration their emotional and spiritual sensibilities. Massive public education programs are needed to promote respect for diversity and the oneness that underlies that diversity. It is less expensive and more effective to attack the root causes of unrest than to stop explosions of violence.

Early warning systems of governments and UN agencies could better connect with NGOs and the media to help generate the political will to prevent or reduce conflicts. User-initiated collaborations on the Web should be increasingly used for peace promotion, rumor control, fact-finding, and reconciliation. Backcasted peace scenarios should be created through participatory processes to show plausible alternatives to conflict stories (see CD Chapter 3.7). It is still necessary, however, to bring to justice those responsible for war crimes and to support the International Criminal Court. The Geneva Convention should be modified to cover intra-state conflicts. Some believe that the collective mind of humanity can contribute to peace or conflict, and hence we can think ourselves into a more peaceful future. Meanwhile, governments should destroy existing stockpiles of biological weapons, create tracking systems for potential bioweapons, establish an international audit system for each weapon type, and increase the use of non-lethal weapons to reduce future revenge cycles. Networks of CDC-like centers to counter impacts of bioterrorism should be also supported. Challenge 10 will be addressed seriously when arms sales and violent crimes decrease by 50% from their peak.

REGIONAL CONSIDERATIONS

AFRICA: The Tunisian and Egyptian revolutions and Libyan internal fighting open North Africa and the wider Arab world to a variety of scenarios. Some believe the death of Osama bin-Laden decreases Al Qaeda's role from Mauritania to Somalia, while others see a rising Muslim Brotherhood. Sub-Saharan Africa has slowly decreased conflicts over the past 10 years. South Sudan has achieved independence. In 2010 there were more than 250,000 Ethiopian IDPs, and there are 300,000 Somali refugees in a Kenyan camp. Serious unrest has broken out between Christians and Muslims in Nigeria, where $22 billion in oil revenues has vanished into local treasuries. Youth unemployment and millions of AIDS orphans may fuel a new generation of violence and crime.

ASIA AND OCEANIA: The popular uprisings have spread from North Africa to Syria, Bahrain, and Yemen. With the potential for the collapse of Yemen, oil piracy along the Somali coast could increase. An internationally acceptable solution to Iran and North Korea's nuclear ambitions is still lacking, and Pakistan's internal instability and uncertain relationships with India and Afghanistan hinder the peacemaking and counter-extremist efforts in all three countries. The $7.5 billion in civilian aid given to Pakistan over the past five years has been largely ineffective. There is no clarity on a NATO withdrawal from Afghanistan, and Iraq's future stability is in doubt. India is facing spreading Maoist violence. Muslim populations from Chechnya to the Philippines are struggling for political and religious rights. Young Palestinians are using online social networks to form a movement separate from Hamas and Fatah to promote the vision of a future Palestinian state. Kurdish aspirations are still a cause of unrest in Turkey and Iraq, but 300,000 Kurds received Syrian citizenship. Relations between North and South Korea have deteriorated. China's internal problems over water, energy, demographics, urbanization, income gaps, and secessionist Muslims in the northwest will have to be well-managed to prevent future conflicts, while tensions with Taiwan are easing.

EUROPE: The EU has created a unit of the External Action Service to actively prevent conflicts. Poland, joined by the Czech Republic, Slovakia, and Hungary, has set up the Visegrad Battle Group, a mini-analogue to NATO. The large numbers of migrant laborers entering the EU will require new approaches to integrate them better into society if increased conflicts are to be prevented. This is aggravated by the new surge of immigrants from the Arab uprisings that Italy has taken in but other countries are unwilling to accommodate. The Roma population continues to be a challenge across the continent.

LATIN AMERICA: Although national wars are rare in the region, internal violence from organized crime paramilitaries continues to be fueled in some areas by corrupt government officials, military, police, and national and international corporations. Mexico's war against organized crime has accelerated, with 35,000 deaths over four years (10,000 of them in 2010). Recent political changes have begun to improve opportunities for indigenous peoples in some parts of the region, while political polarization over policies to address poverty and development persists. Colombia plans on returning to the rightful owners 2.5 million hectares of land seized by gangs. Violence is impeding development in Central America.

NORTH AMERICA: As Arctic ice continues to melt, vast quantities of natural gas and oil will be accessible where national boundaries are under dispute. This could be a source of U.S.-Canadian tension, along with Russia, Norway, and Denmark. The U.S. Institute of Peace's SENSE multi-person training simulation has educated thousands of participants worldwide in the fundamentals of decisionmaking, resource allocation, and negotiation in post-conflict situations. Cooperation on environmental security could become a focus of U.S.-China strategic trust.

11. How can the changing status of women help improve the human condition?

Empowerment of women has been one of the strongest drivers of social evolution over the past century, and many argue that it is the most efficient strategy for addressing the global challenges in this chapter. Only two countries allowed women to vote at the beginning of the twentieth century; today there is virtually universal suffrage, the average ratio of women legislators worldwide has reached 19.2%, and over 20 countries have women heads of state or government. Patriarchal structures are increasingly challenged, and the movement toward gender equality is irreversible.

With an estimated control of over 70% of global consumer spending, women are strongly influencing market preferences. Analysis shows a direct interdependence between countries' Gender Gap Index and their Competitiveness Index scores and that Fortune 500 companies with more gender-balanced boards could outperform the others by as much as 50%. Yet the Gender Equity Index 2010 shows that significant differences still remain in economic participation and political empowerment.

Gender stereotyping continues to have negative impacts on women around the world, and although progress is being made on closing the gender gap in terms of establishing global and national policies, real improvement will only be achieved when conflicts between written laws and customary and religious laws and practices are eliminated. Environmental disasters, food and financial crises, armed conflicts, and forced displacement further increase vulnerabilities and generate new forms of disadvantages for women and children.

Women account for over 40% of the world's workforce, earn less than 25% of the wages, and represent about 70% of people living in poverty. An OECD survey found that women spend more time on unpaid work than men do worldwide, with the gap ranging from 1 hour per day in Denmark to 5 hours per day in India. FAO estimates that giving women the same access as men to agricultural resources could reduce the number of hungry people in the world by 12–17%, or 100–150 million people. Child malnutrition levels are estimated to be 60% above average where women lack the right to land ownership and 85% above average where they have no access to credit. Microcredit institutions reported that by 2010, nearly 82% (about 105 million) of their poorest clients were women. However, many of their businesses are too small to transform their economic status, points out FEMNET.

Empowerment of women is highly accelerated by the closing gender gap in education. Most countries are reaching gender parity in primary education, and

50% of university students worldwide are women. Yet regional disparities are high, and UNESCO estimates that women represent about 66% of the 796 million adults who lack basic literacy skills.

Although the health gender gap is closing, family planning and maternal health remain critical. Determining the size of the family should be recognized as a basic human right, and more attention should be given to women's health and social support for affordable child care worldwide, including industrial countries, which are facing demographic crises due to low fertility rates. Of the more than 500,000 maternal deaths per year, 99% happen in developing countries, with the highest prevalence in Africa and Asia due to high fertility rates and weak health care systems. Unless providing effective family planning to the 215 million women who lack it is seen as a key component of development, the UN goal to reduce maternal mortality to 120 deaths per 100,000 live births by 2015 will not be achieved.

Regulations should be enacted and enforced to stop female genital mutilation, which traumatizes about 3 million girls in Africa each year, in addition to the 100–142 million women worldwide affected by it today. While the prevalence of this in Egypt, Guinea, and some parts of Uganda is at over 90%, communities in India, Indonesia, Malaysia, and even in the EU are also affected.

Violence against women is the largest war today, as measured by death and casualties per year. While the proportion of women exposed to physical violence in their lifetime ranges from 12% to 59%, a function of region and culture, sexual assaults remain one of the most underreported crimes worldwide, continuing to be perpetrated with impunity.

According to UNODC, 66% of the victims of the $32 billion global industry of human trafficking are women and children. The Protocol to Prevent, Suppress and Punish Trafficking in Persons, especially Women and Children, has 142 parties and 117 signatories thus far, but it has yet to be adopted and enforced by some key countries.

Female vulnerability increases during conflict, when sexual violence is often used as a weapon. Recovery from conflict and disaster should be used as opportunities to rectify inequalities. Nevertheless, women make up only 8% of peace negotiators, and only 25 countries have developed National Action Plans supporting UN Security Council Resolution 1325 on women's protection in conflict and participation in peace processes.

Traditional media have had limited success influencing gender stereotyping, and women represent only one-third of full-time workers in journalism, reveals an IWMF survey. However, 78% of women (versus 66% of men) are active users of social media, a new powerful medium for change.

Mothers should use their educational role in the family to more assertively nurture gender equality. School systems should consider teaching self-defense in physical education classes for girls. Infringements on women's rights should be subject to prosecution and international sanctions. (See Appendix N in the CD for an annotated list of resources addressing gender equity and the study conducted by Millennia 2015 on potential policies to improve the status of women.)

Challenge 11 will be addressed seriously when gender-discriminatory laws are gone, when discrimination and violence against women are prosecuted, and when the goal of 30%+ women's representation in national legislatures is achieved in all countries.

REGIONAL CONSIDERATIONS

AFRICA: Despite significant progress with enrolment in primary education, dropout rates are 40% and one in three children is engaged in child labor. Half of the world's maternal deaths occur in sub-Saharan Africa, and women have little say in their own health care. African countries experiencing conflict or natural disasters have a very high incidence of sexual violence. Women represent 19.1% of legislatures in sub-Saharan African, with Rwanda being the world's only women-majority parliament. However, conflicts between constitutional and customary law are an issue across the region, and only 29 of the 53 African Union countries ratified the Protocol on Women's Rights.

ASIA AND OCEANIA: The average lifetime risk of maternal death in rural South and East Asia and the Pacific is over 4%. The preference for male children, largely due to inheritance laws and dowry liabilities, is causing a gender imbalance in many countries in the region, most notably in China and India where, in some communities, the birth ratio is as low as 60–70 females to every 100 males. According to UNICEF, child marriage is a severe issue in Nepal and some parts of India, where about 40% of girls become child brides. Women's representation in Arab States' legislature reached 10.7% (from 3.6% in 2000), the average of economically active adult female rose to 28%, and the social uprisings of 2011 are expected to further unsettle the patriarchal society dominating most Muslim-majority countries. In places such as

Afghanistan, where 85% of women are illiterate and only 37% of students in schools are girls, women's rights should be central to peace and development agreements. Half of the world's top 20 richest self-made women are Chinese.

EUROPE: Women hold 41.6% of parliamentary seats in Nordic countries, 20.8% in OSCE countries (excluding Nordic ones), and 35.2% of EU Parliament seats. The proportion of women on the boards of the top European companies was 12% in 2010; at current rates, this would reach parity in 16 years. Women make up 7% of the directors in Russia's 48 biggest companies, and a new draft law proposes at least 30% of parliamentary seats are to be occupied by women, as well as providing advantages for men to play a greater role in family life. The UN estimates that there are between 200,000 and 500,000 illegal sex workers in the EU, the majority from Central and Eastern Europe. France's rule banning full-face veils in public is aiming to enforce women's rights and might be emulated by other EU countries.

LATIN AMERICA: Women's participation in Latin American parliaments improved due to the introduction of quotas in many countries, while Argentina, Brazil, and Costa Rica have female presidents. More women than men attain tertiary education across the region, but wage discrepancies persist. Although all countries in the region have ratified the Convention on the Elimination of all Forms of Discrimination against Women, as a result of restrictive legislation, one in three maternal deaths is due to abortion and the lifetime risk of maternal death is 0.4%. Mexico's programs and actions for the rights of women are among the best practices in Latin America, and many countries in the region are interested in emulating them.

NORTH AMERICA: Women make up half of the U.S. workforce, with an unemployment rate 1% lower than men's; more women than men are gaining advanced college and bachelor's degrees, redefining the roles in the family. Nevertheless, although they hold 51.5% of management, professional, and related positions, women account for only 3% of the Fortune 500 chief executives. The U.S. government estimates that approximately 75% of the 14,500–17,500 people annually trafficked into the country are female. Women's representation in U.S. legislature is 16.9%, while Canada's is 25%. Both U.S. and Canadian governments made critical cuts in domestic and international family planning programs for women.

12. How can transnational organized crime networks be stopped from becoming more powerful and sophisticated global enterprises?

Although the world is waking up to the enormity of the threat of transnational organized crime, it continues to grow, while a global strategy to address this global threat has not been adopted. The UN Office on Drugs and Crime has called on all states to develop national strategies to counter TOC as a whole in its report *The Globalization of Crime: A Transnational Organized Crime Threat Assessment*. The transition of much of the world's activities to the Internet and mobile phones has opened up a wealth of opportunities for TOC to profitably expand its activities from drugs and human trafficking to all aspects of personal and business life. The financial crisis and the bankruptcy of financial institutions have opened new infiltration routes for TOC crime, and the world recession has increased human trafficking and smuggling. UNODC also notes states are not seriously implementing the UN Convention against Transnational Organized Crime, which is the main international instrument to counter organized crime. INTERPOL has started construction of an international complex in Singapore that will open around 2013 with 300 staff, to serve as a center for policy, research, and worldwide operations. UNODC, with other agencies, has founded the International Anti-Corruption Academy, near Vienna, having as one of its goals tackling the connection of TOC and corruption. That combination, allowing government decisions to be bought and sold like heroin, makes democracy an illusion.

Havocscope.com estimates world illicit trade to be about $1.6 trillion per year (up $500 billion from last year), with counterfeiting and intellectual property piracy accounting for $300 billion to $1 trillion, the global drug trade at $404 billion, trade in environmental goods at $63 billion, human trafficking and prostitution at $220 billion, smuggling at $94 billion, and weapons trade at $12 billion. The International Carder's Alliance is based mostly in Eastern Europe, the heart of cybercrime, which the FBI estimates costs U.S. businesses and consumers billions annually in lost revenue. These figures do not include extortion or organized crime's part of the $1 trillion in bribes that the World Bank estimates are paid annually or its part of the estimated $1.5–6.5 trillion in laundered money. Hence the total income could be $2–3 trillion—about twice as big as all the military budgets in the world. The UN Global Commission on Drug Policy concluded that the law enforcement of the "War on Drugs" has failed and cost the U.S. $2.5 trillion over 40 years. It recommends a "paradigm shift" to public health over criminalization. OECD's Financial Action Task Force has made 40 recommendations to counter money laundering.

There are more slaves today than at the peak of the African slave trade. Estimates range from 12 million to 27 million people are being held in slavery today (the vast majority in Asia). UNICEF estimates that 1.2 million children are trafficked every year. The online market in illegally obtained data and tools for committing data theft and other cybercrimes continues to grow, and criminal organizations are offering online hosting of illegal applications. International financial transfers via computers of $2 trillion per day make tempting targets for international cyber criminals.

It is time for an international campaign by all sectors of society to develop a global consensus for action against TOC. Two conventions help bring some coherence to addressing TOC: the UN Convention against Transnational Organized Crime, which came into force in 2003, and the Council of Europe's Convention on Laundering, which came into force in May 2008. Possibly an addition to one of these conventions or the International Criminal Court could establish a financial prosecution system as a new body to complement the related organizations addressing various parts of TOC. In cooperation with these organizations, the new system would identify and establish priorities on top criminals (defined by the amount of money laundered) to be prosecuted one at a time. It would prepare legal cases, identify suspects' assets that can be frozen, establish the current location of the suspect, assess the local authorities' ability to make an arrest, and send the case to one of a number of preselected courts. Such courts, like UN peacekeeping forces, could be identified before being called into action and trained, and then be ready for instant duty. When all these conditions are met, then all the orders would be executed at the same time to apprehend the criminal, freeze access to the assets, open the court case, and then proceed to the next TOC leader on the priority list. Prosecution would be outside the accused's country. Although extradition is accepted by the UN Convention against Transnational Organized Crime, a new protocol would be necessary for courts to be deputized like military forces for UN peacekeeping, via a lottery system among volunteer countries. After initial government funding, the system would receive its financial support from frozen assets of convicted criminals rather than depending on government contributions.

Challenge 12 will be seriously addressed when

money laundering and crime income sources drop by 75% from their peak.

REGIONAL CONSIDERATIONS

AFRICA: The West Africa Coast Initiative is a partnership among UNODC, UN Peacekeeping, ECOWAS, INTERPOL, and others to address the problems that have allowed traffickers to operate in a climate of impunity. The drug traffic from Latin America through the West African coast to Africa and Europe has been declining. Piracy, centering on Somalia and now invading the Indian Ocean, has become a major crisis, with 286 piracy incidents worldwide in 2010 and 67 hijacked ships and over 1,130 seafarers affected. Yemen's unsettled future could leave the oil shipping lanes of the Arabian Sea bordered by two failed states. A multinational naval force is combating the problem, which is complicated by the lack of clear international legal structures for prosecution and punishment; Kenya has withdrawn its support in this area, but Somaliland and Puntland are helping. Some 930,000 sailors have signed a petition to the IMO to stop piracy. Smoking of "whoonga" has become a serious problem in South Africa, with robberies committed to support £90 per day habits. The 15 million AIDS orphans in sub-Saharan Africa, with few legal means to make a living, constitute a gigantic pool of new talent for the future of organized crime. Corruption remains a serious impediment to economic development in many African countries.

ASIA AND OCEANIA: The International Conference on Asian Organized Crime and Terrorism held its eighth meeting to share intelligence. Disease cut Afghan opium production by almost half, but the acreage stayed the same; four more provinces became almost drug-free, but domestic addiction is spreading. China is the main source for counterfeit goods sent to the EU. India is a major producer of counterfeit medicines. North Korea is perceived as an organized crime state backed up by nuclear weapons involved in illegal trade in weapons, counterfeit currency, sex slavery, drugs, and a range of counterfeit items. Myanmar is accused of deporting migrants to Thailand and Malaysia, where they are exploited, and has reportedly become a center for the ivory trade and elephant smuggling. Myanmar and China remain the primary sources of amphetamine-type stimulants in Asia; Myanmar rebels are exporting hundreds of millions of tablets to Thailand to raise money. A report says Australia has a multi-billion-dollar drug enterprise, and Australians are among the world's highest per capita consumers of illicit stimulants.

EUROPE: Europol published a 2011 TOC Threat Assessment at www.europol.europa.eu, indicating greater TOC mobility, operational diversity, and internal collaboration. The EU has strengthened controls on money transfers across its borders to address trafficking and money laundering, especially in Eastern Europe. Russian officials have declared the drug situation in that country "apocalyptic." An estimated €30 million in EU carbon emission allowances were stolen in January. The Italian Guardia di Finanza arrested 24 people from a €2.7 billion Chinese counterfeit fashion operation, and Milan police arrested 300 members of 'Ndrangheta. London police smashed a £100 million drug and money laundering gang.

LATIN AMERICA: About 35,000 people have died in the Mexican drug war over the last four years, of which 15,000 died in 2010. Mexico's cartels receive more money (an estimated $25–40 billion) from smuggling drugs to the U.S. than Mexico earns from oil exports. About $1 billion worth of oil was stolen from pipelines (396 taps) and smuggled into the U.S. over a two-year period. Mexican drug cartels are rapidly moving south, into Central America, and are branching out, with La Familia exporting $42 million worth of stolen iron ore from Michoacán in a year. UNODC says crime is the single largest issue impeding Central American stability. Cocaine production in Colombia has dropped by two-thirds and is now done by small gangs, using farms hidden in the jungles. The drug gangs have largely replaced the paramilitaries. Ecuador has become an important center of operation for TOC gangs (3,000 people have reportedly moved in from Colombia). Police seized a drug smuggling submarine in Colombia. Drug cartels exist in Latin America because of illegal drug consumption in the U.S.

NORTH AMERICA: The International Organized Crime Intelligence and Operations Center integrates U.S. efforts to combat international organized crime and coordinates investigations and prosecutions. The 22-month anti-cross-border-drug Project Deliverance ended successfully in June 2010 after the arrest of 2,200 individuals and the seizure of more than 69 tons of marijuana, 2.5 tons of cocaine, 1,410 pounds of heroin, and $154 million in currency. Only about 53 miles of operational "virtual fence" were put in place in Arizona, at a cost of about $15 million a mile; the project has been dropped. Drug criminal gangs have swelled in the U.S. to an estimated 1 million members, responsible for up to 80% of crimes in communities across the nation. Organized crime and its relationship to terrorism should be treated as a national security threat. Canada continues as a major producer and shipper of methamphetamines and ecstasy.

CHAPTER ONE

13. How can growing energy demands be met safely and efficiently?

Investments into alternatives to fossil fuels are rapidly accelerating around the world to meet the projected 40–50% increase in demand by 2035. The combined global installed capacity of wind turbines, biomass and waste-to-energy plants, and solar power reached 381 GW, exceeding the installed nuclear capacity of 375 GW (figure prior to the Fukushima disaster). The Japanese nuclear disaster has put the future of nuclear energy in doubt, increasing costly safety requirements and reducing public and investor confidence. This, plus the BP oil disaster and the growing awareness of climate change, are accelerating the transition to renewable energy sources. However, without major breakthroughs in technological and behavioral changes, the majority of the world's energy in 2050 will still come from fossil fuels. Therefore, large-scale carbon capture and reuse has to become a top priority to reduce climate change, such as using waste CO_2 from coal plants to grow algae for biofuels and fish food or to produce carbonate for cement. The short-term gainer may be natural gas. Energy efficiencies, conservation, and reduced meat consumption are near-term ways to reduce energy GHG production. To keep atmospheric CO_2 concentration below 450 ppm, an estimated $18 trillion investment on low-carbon technologies will be needed between 2010 and 2035. Meanwhile, the world spends more than $310 billion on energy subsidies every year; eliminating these could reduce GHGs by 10% by 2050.

Global investments in clean energy reached $243 billion in 2010, up from $186.5 billion in 2009. China leads the world in total investments in renewable energy and energy efficiency. IPCC's best-case scenario estimates that renewable sources could meet 77% of global energy demand by 2050, while WWF claims 100% is possible. Setting a price for carbon emissions will stimulate investments. For the past decade, coal has met 47% of new electricity demand globally. Assuming that countries fulfill their existing commitments to reduce emissions and cut fuel subsidies, IEA estimates that the world primary energy demand will still increase by 36% from 2008 to 2035, or 1.2% per year, with fossil fuels accounting for over half of the increase. World energy consumption increased 5% in 2010 after shrinking 1.1% in 2009.

IEA says $36 billion/year will connect the remaining 1.4 billion people around the world with electricity. About 3 billion people still rely on traditional biomass for cooking and heating, and 1.4 million people die every year due to indoor smoke from traditional cookstoves. The UN has declared 2012 as the International Year of Sustainable Energy and set 2030 for universal access to modern energy sources.

Auto manufacturers around the world are racing to create alternatives to petroleum-powered cars. Mass production of fuel-flexible plug-in hybrid electric cars at competitive prices could be a breakthrough in de-carbonizing the transport sector. The global share of biofuel in total transport fuel could grow from 3% today to 27% in 2050. Massive saltwater irrigation along the deserted coastlines of the world can produce 7,600 liters/hectare-year of biofuels via halophyte plants and 200,000 liters/hectare-year via algae and cyanobacteria, instead of using less-efficient freshwater biofuel production that has catastrophic effects on food supply and prices. Drilling and liquefaction of natural gas via integrated ships promises to get more LNG in less time and cost.

Innovations are accelerating: concentrator photovoltaics to dramatically reduce costs; pumping water through micro-channels on the surface of a solar panel to make it more efficient and make seawater drinkable at the same time; producing electricity from waste heat from power plants, human bodies, and microchips; genomics to create hydrogen-producing photosynthesis; buildings to produce more energy than consumed; solar energy to produce hydrogen; microbial fuel cells to generate electricity; and compact fluorescent light bulbs and light-emitting diodes to significantly conserve energy, which can also be done by nanotubes that conduct electricity. Solar farms can focus sunlight atop towers with Stirling engines and other generators. Estimates for the potential of wind energy continue to increase, but so do maintenance problems. Drilling to hot rock (two to five kilometers down) could make geothermal energy available where conventional geothermal has not been possible. Plastic nanotech photovoltaics printed on buildings and other surfaces could cut costs and increase efficiency. The transition to a hydrogen infrastructure may be too expensive and too late to affect climate change, while flex-fuel plug-in hybrids, electric, and compressed air vehicles could provide alternatives to petroleum-only vehicles sooner. Unused nighttime power production could supply electric and plug-in hybrid cars. National unique all-electric car programs are being implemented in Denmark and Israel, with discussions being held in 30 other countries. Behavior changes and conservation can reduce demand.

Japan plans to have a working space solar power system in orbit by 2030. Such space-based solar energy systems could meet the world's electricity requirements indefinitely without nuclear waste or GHG emissions.

Eventually, such a system of satellites could manage base-load electricity on a global basis, yet some say this costs too much and is not necessary with all the other innovations coming up.

Challenge 13 will have been addressed seriously when the total energy production from environmentally benign processes surpasses other sources for five years in a row and when atmospheric CO_2 additions drop for at least five years.

REGIONAL CONSIDERATIONS

AFRICA: Over 70% of sub-Saharan Africa does not have access to electricity. The World Bank's Lighting Africa initiative mobilizes funding from the private sector to provide affordable and modern off-grid lighting to 2.5 million people in Africa by 2012 and to 250 million people by 2030. The $80 billion Grand Inga dam could generate 40,000 MW of electricity, but the project is progressing slowly due to political instability, mismanagement of public finance, and possible environmental and social impacts. Algeria will invest $60 billion in renewable energy projects by 2030. By 2050, some 10–25% of Europe's electricity needs could be met by North African solar thermal plants.

ASIA AND OCEANIA: There are more people without electricity in India (400 million) than live in the U.S. China uses more coal than the U.S., Europe, and Japan combined; it also builds more-efficient, less-polluting coal power plants. It is expected to add generation capacity equivalent to the current total installed capacity of the U.S. in the next 15 years. China invested more than $64 billion (1.4% of its GDP) in clean energy in 2010 and plans to expand its offshore wind turbines to 5 GW by 2015 and 30 GW by 2020. It added nearly 20 million vehicles in 2010 and now produces more cars than the U.S. and Japan, and it could lead the world in electric car production. India will invest $37 billion in renewable energy to add additional capacity of 17,000 MW by 2017. Oil and gas production in the Caspian region will grow substantially in the next 20 years; Kazakhstan and Turkmenistan lead the growth in oil and gas respectively. China has 13 nuclear reactors in operation and 25 under construction. India plans to increase nuclear energy's share from 3% to 13% by 2030.

EUROPE: Conservation and efficiencies could reduce EU's energy consumption about 30% below 2005 levels by 2050. Low-carbon technologies could provide 60% of energy by 2020 and 100% by 2050 according to the EU's low carbon roadmap. Germany and Switzerland plan to phase out nuclear energy. Increasing imports of renewable energy from MENA and natural gas from Eastern Europe seem inevitable. The future pan-European smart grid should allow massive deployment of low-carbon energy supply. A Swedish team certified Italian claims that low-energy nuclear reactions produced sufficiently more energy than consumed over 18 hours to trigger commercial planning. EU plans to have 10–12 carbon capture and storage demonstration plants in operation by 2015. Amsterdam plans to have 10,000 electric cars by 2015. Five geothermal power plants in Iceland supply 27% of the country's electricity needs. Europe is on track to generate 20% of its energy from renewable sources by 2020.

LATIN AMERICA: Brazil is the world's second largest producer of bioethanol, with 33% of the world market, producing it at 60¢ per gallon and meeting 40% of its automotive needs; 90% of the automobiles produced in Brazil are flex-fuel. Argentina is the world's second largest producer of biodiesel, with 13.1% of the market. Geothermal, solar, and wind are vast untapped resources for the region, as are gains from efficiencies. Installed wind power capacity in the region is expected to grow by 12.6% per year and reach 46 GW by 2025, with Brazil and Mexico having a dominant share. Ecuador announced that it would refrain from drilling for oil in the Amazon rainforest reserve in return for up to $3.6 billion in payments from industrial countries. Venezuela's Orinoco heavy oil reserves (requiring advanced production technology) are larger than Saudi Arabia's reserves. Argentina, Brazil, and Mexico have nuclear reactors but have not changed their nuclear policy, while Venezuela froze its plan to develop nuclear energy.

NORTH AMERICA: Lesser-known potential clean energy sources in the U.S. include high-altitude wind off the East Coast, OTEC in the Gulf Stream, solar thermal in the Midwest (four corners), drilled hot rock geothermal, and nano-photovoltaics. The U.S. investment in clean energy increased by 51% in 2010, but the U.S. dropped to third place after China and Germany. Algae farms for biofuel may cost $46.2 billion per year to replace oil imports. California requires oil refineries and importers of motor fuels to reduce the carbon intensity of their products by 10% by 2020. San Francisco's mayor called for the city to go 100% renewable by 2020. Pacific Gas & Electric Company of California agreed to buy 200 megawatts of space-based solar power by 2016 from Solaren. Recycling waste heat from nuclear power plants to home air conditioners and recycling body heat to recharge batteries could reduce CO_2 by 10–20% in the U.S.

14. How can scientific and technological breakthroughs be accelerated to improve the human condition?

The acceleration of S&T continues to fundamentally change the prospects for civilization, and access to its knowledge is becoming universal. The ability to learn this knowledge is also improving with Web-based asynchronous highly motivational educational systems, adaptive learning models such as cellular automata, genetic algorithms, neural networks, and emerging capabilities of collective intelligence systems. Computing power and lowered costs predicted by Moore's Law continues with the world's first three-dimensional computer chip introduced by Intel for mass production. Computational chemistry, computational biology, and computational physics are changing the nature of science, and its acceleration is attached to Moore's law. China currently holds the record for the fastest computer with Tianhe-1, which can perform 2.5 petaflops per second; IBM's Mira, ready next year, will be four times faster. Watson is the IBM computer that beat the top knowledge contestants on a TV quiz show; it is a massively parallel processing computer capable of reading an essentially unlimited number of documents, digesting the information, and answering questions posed in natural language. It is now being readied to use vast amounts of medical data to accelerate improvements in health knowledge and decisionmaking. Watson supports the idea that intelligent computer systems can be smarter than humans.

Craig Venter created a synthetic genome by placing a long strand of synthetic DNA into a bacterium that followed the synthetic DNA's instructions and replicated. A U.S. Presidential Commission concluded that it was not yet the invention of "life" but that synthetic biology research should continue with scientific self-regulation. Venter forecasts that as computer code is written to create software to augment human capabilities, so too genetic code will be written to create life forms to augment civilization. In a process known as transdifferentiation, scientists have manipulated human cells, converting pancreatic cells into liver cells and skin cells into heart cells; skin cells were converted into functioning neurons that could integrate into neuron networks of the sort found in the human brain. A new anti-virus strategy is being pursued to develop artificial "proto-cells that can lure, entrap and inactivate a class of deadly human viruses."

Nano robots now roam inside the eyes in tests to deliver drugs for conditions such as age-related macular degeneration. Swarms of manufacturing robots are being developed that should be able to manage nano-scale building blocks for novel material synthesis and structures, component assembly, and self-replication and repair. At an even smaller scale, nanometer robots have been demonstrated and appear able to link with natural DNA. Nanobots the size of blood cells may one day enter the body to diagnose and provide therapies and internal virtual reality imagery. Although nanotech promises to make extraordinary gains in efficiencies needed for sustainable development, its environmental health impacts are in question.

Scanning electron microscopes can see 0.01 nanometers (the distance between a hydrogen nucleus and its electron), and the Hubble telescope has seen 13.2 billion light-years away. Photons have been slowed and accelerated. External light has been concentrated inside the body for photodynamic therapy and powered implanted devices. DNA scans open the possibility of customized medicine and eliminating inherited diseases. MRI brain imaging shows primitive pictures of real-time thought processes. Paralyzed people have controlled computers with their thoughts alone.

Anti-matter has been trapped (in the form of 309 atoms of antihydrogen) in electro-magnetic containment and observed for an astonishing 17 minutes in CERN's particle physics laboratory. This may facilitate research into how gravity and time affect antimatter. Some scientists predict that if the Large Hadron Collider succeeds in producing the Higgs boson, it may also create a second particle called the Higgs singlet that should have the ability to jump into an extra, fifth dimension where they can move either forward or backward in time and reappear in the future or past. On another frontier one group is attempting to entangle billions of particle pairs (quantum entanglement is the simultaneous change of entangled objects separated in space). Quantum building blocks, qubits, have been embedded into nanowires, important steps toward quantum computers. Quantum theory also encompasses the "many worlds interpretation" of our existence. In the MWI, every event is a branch point that may go this way or that, creating an almost infinite set of branches. Follow any one and it describes a simultaneously existing alternate world, a remarkable and counterintuitive reality. Although seemingly remote from improving the human condition, such basic science is necessary to increase knowledge that applied science and technology draws on to improve the human condition.

We need a global collective intelligence system to track S&T advances, forecast consequences, and document a range of views so that politicians and the public can understand the potential consequences of new S&T. Challenge 14 will have been addressed seriously when the funding of R&D for societal needs reaches parity with funding for weapons and when an international science and technology organization is established that routinely connects world S&T knowledge for use in R&D priority setting and legislation.

REGIONAL CONSIDERATIONS

AFRICA: The first Inter-Parliamentary Forum on Science, Technology and Innovation promises to increase the percent of GDP for S&T. African Innovation Outlook 2010 found S&T for medicine has passed agriculture, but Africa's share of global science continues to decrease. These low levels of R&D investment, weak institutions, and poor access to markets are among the key challenges in actualizing Africa's innovation potential. The UN Economic Commission for Africa is supporting science training via collaboratories to connect African scientists with counterparts overseas to use S&T more efficiently. Primary commodities account for 80% of Africa's exports; S&T innovation is needed to create added value exports and to leapfrog into future biotechnology, nanotech, and renewable energy prospects. UNESCO's 2010 World Social Science Report found a 112% increase in Africa's publications in social studies and humanities between 1987 and 2007.

ASIA AND OCEANIA: Chinese patent filings have gone up 500% in the last five years; China is investing more in cleaner energy technology than the U.S. does and it has the second largest R&D budget in the world. Asian countries with double-digit economic growth also have double-digit growth in R&D expenditures. Energy and environment is the focus of U.S. and China relations. Japan has launched a Venus probe that also carried a space sail that gains its energy from solar "wind" pressure in space.

EUROPE: The 2012 EU budget increases research by 13%. The EU is establishing a single European system for registering patents. Although the Lisbon Strategy expired in 2010, succeeded by Europe 2020, the EU target of 3% of GDP for R&D has been kept. Only two EU member states have achieved the 3% target so far, while the average R&D expenditure of the EU27 stood at 2.01% of GDP in 2009. The newer members' R&D expenditure remains low, with many under 1%. Russia has lost over 500,000 scientists over the last 15 years, but a reverse trend is beginning, salaries have increased, innovation is encouraged, and high tech is being supported. Russian investments in nanotechnology R&D and corporations have been substantial, even during the recent recession. Russia is building the Skolkovo Innovation Center with funding from multinational corporations to accelerate R&D and applications.

LATIN AMERICA: OECD, UNESCO, EU, the U.S., and China are helping countries in the region with innovation systems. Chile has started a scientific news network for Latin America in order to reverse some of the lagging indicators in the region. Argentina, Brazil, Chile, and Mexico account for almost 90% of university science in the region, and half of the 500 higher education institutes produce no scientific research. University S&T courses could be required to focus some attention on helping the poorest communities. Mexico is leading the Innovation Network for Latin American and the Caribbean.

NORTH AMERICA: Research by the U.S. National Academy of Sciences, National Academy of Engineering, and Institute of Medicine is available for free downloads. The Massachusetts Institute of Technology makes 2,000 courses online–many the top S&T course in the world–available at no cost with videos, lecture notes, and references. The U.S. Peace Corps has created Information Volunteers to help developing countries access science and technology information in the classroom. The space shuttles made their last flights in 2011. About 35% of world R&D is in the U.S. Each week the U.S. Patent Office makes thousands of new patents freely available online. Prizes can speed the distribution of technology that benefits humanity, such as the Tech Awards from the Tech Museum in San Jose, California, or Richard Branson's new prize for a plan to remove a billion tons of carbon dioxide a year, as can tech sports like MIT's robot competitions.

Figure 7: Patents issued in the U.S. (per year)

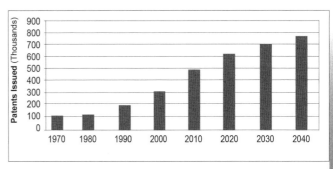

Source: U.S. Patent Statistics

15. How can ethical considerations become more routinely incorporated into global decisions?

Is the acceleration of global change beyond conventional means of ethical evaluation? Must we invent anticipatory ethical systems? Just as law has a body of previous judgments upon which to draw for guidance, will we also need bodies of ethical judgments about future possible events? For example, is it ethical to clone ourselves or bring dinosaurs back to life or invent thousands of new life forms from synthetic biology? Despite the extraordinary achievements of S&T, future risks from their continued acceleration and globalization remain (see CD Chapter 3.5) and give rise to future ethical issues (see CD Chapters 5 and 11). On the brighter side, new technologies also make it easier for more people to do more good at a faster pace than ever before. Single individuals initiate groups on the Internet, organizing actions worldwide around specific ethical issues. News media, blogs, mobile phone cameras, ethics commissions, and NGOs are increasingly exposing unethical decisions and corrupt practices.

The killing of bin-Laden raises political, legal, and ethical issues with a broad range of views. Some examples: it may have improved the future by hastening the collapse of Al Qaida; the U.S. has killed more innocent people in response to the smaller number killed on 9/11; it will prevent many future deaths; vengeance is not justice. Related to this are the future possibilities and ethical issues of a single individual who is massively destructive. Hopefully a future SIMAD is identified before he or she has the chance to be massively destructive. Picture the person building new bio-viruses in a basement laboratory that if delivered could kill millions of people. Do civil rights apply, or should society impose sanctions before the fact? In bin-Laden's case, was the killing justified because many believed it will help save lives and create a better future? To reduce the number of future SIMADs, healthy psychological development of all children should be the concern of everyone. Such observations are not new, but the consequences of failure to realize their importance may be much more serious in the future than in the past.

The moral will to act in collaboration across national, institutional, religious, and ideological boundaries that is necessary to address today's global challenges requires global ethics. Public morality based on religious metaphysics is challenged daily by growing secularism, leaving many unsure about the moral basis for decisionmaking. Unfortunately, religions and ideologies that claim moral superiority give rise to "we-they" splits.

The UN Global Compact—with 8,000 participants, including over 5,300 businesses in 130 countries—was created to reinforce ethics in decisionmaking; it has improved business-NGO collaboration, raised the profile of corporate responsibility programs, and increased businesses' non-financial reporting mandates in many countries. The Compact has been used to encourage corporations to urge their countries to ratify the UN Convention against Corruption, which has been ratified by 143 states. As of March 2011 there were 26 first-year country reviews of corruption under way via the convention. Article 51 calls on states parties to return stolen assets; the unethically acquired wealth by Arab dictators is being uncovered, and this might be a test of this article. The ICC has successfully tried political leaders, and proceedings are Web-cast.

Some believe that Wikileaks will ultimately improve ethical considerations in global decisions, since, it is argued, it shows that many unethical decisions led to poorer results than expected. The global financial crisis demonstrated the interdependence of economics and ethics.

Although quick fixes have pulled the world out of recession, the underlining ethics has not been addressed sufficiently to prevent future crises. The Universal Declaration of Human Rights continues to shape discussions about global ethics and decisions across religious and ideological divides. UNESCO's Global Ethics Observatory is a set of databases of ethics institutions, teaching, codes of contact, and experts.

Collective responsibility for global ethics in decisionmaking is embryonic but growing. Corporate social responsibility programs, ethical marketing, and social investing are increasing. Global ethics also are emerging around the world through the evolution of ISO standards and international treaties that are defining the norms of civilization. Yet 12–27 million people are slaves today, more than at the height of the nineteenth-century slave trade; the World Bank estimates over $1 trillion is paid each year in bribes; and organized crime takes in $2–3 trillion annually.

Transparency International's Corruption Perception Index measures perceived levels of corruption in the public sector in 178 countries. In 2010 the least corrupt countries were seen to be Denmark, New Zealand, and Singapore; the most corrupt were Equatorial Guinea, Burundi, and Chad. Sixty percent of the people surveyed said that they thought corruption in their countries had increased over the last three years; in Europe, almost 75% said they thought things were getting worse. Around the world, 25% said that they paid bribes in the last year to police (mentioned most often), tax authorities, and other officials. In poor countries, half the people reported paying a bribe in the past year, usually for permits, improved services, and to "avoid problems with authorities."

Entertainment media could promote memes like "make decisions that are good for me, you, and the world." We need to create better incentives for ethics in global decisions, promote parental guidance to establish a sense of values, encourage respect for legitimate authority, support the identification and success of the influence of role models, implement cost-effective strategies for global education for a more enlightened world, and make behavior match the values people say they believe in. Ethical and spiritual education should grow in balance with the new powers given to humanity by technological progress. Challenge 15 will be addressed seriously when corruption decreases by 50% from the World Bank estimates of 2006, when ethical business standards are internationally practiced and regularly audited, when essentially all students receive education in ethics and responsible citizenship, and when there is a general acknowledgment that global ethics transcends religion and nationality.

REGIONAL CONSIDERATIONS

AFRICA: The North African uprisings in 2011 were calls for ethics in decisionmaking. Transparency International chapters in sub-Saharan Africa work to counter corruption. The Business Ethics Network of Africa continues to grow, with conferences, research, and publications. Most African government anti-corruption units are not considered successful, Eight African countries surveyed by Transparency International report that 20% of those interviewed in eight African countries surveyed who had contact with the judicial system reported having paid a bribe.

ASIA AND OCEANIA: As China's global decisionmaking role increases, it will face traditional versus western value conflicts. Some believe the rate of urbanization and economic growth is so fast in Asia that it is difficult to consider global ethics, while Asians do not believe there are common global ethics and maintain that the pursuit to create them is a western notion.

EUROPE: Long-range demographic projections indicate Europe will become the first Moslem-majority content. The European integration processes both help and challenge ethical standards as cultures meet and question each other's way of thinking and acting. Its future immigration policies will have global significance, increasing discussions of ethics and identity for Europe. The European Ethics Network is linking efforts to improve ethical decisionmaking, while Ethics Enterprise is working to mobilize an international network of ethicists and organizes innovative actions to attract attention for ethics in business.

LATIN AMERICA: University courses in business ethics are growing throughout Latin America. Problems such as lack of personal security, limited access to education and health services, lack of faith in politics, badly damaged institutions that do not fulfill their role (such as the Justice system and police), and accelerated environmental degradation in some countries are aspects of a serious lack of ethical values. The prevalence of legal formality, in other countries, does not guarantee equal rights, as large sections of the population remain excluded from the guarantees of goods and people. It also manifests a serious lack of ethical standards in the mass media.

NORTH AMERICA: With the emergence of the G8 and BRIC and the increasing powers of the WTO, ICC, regional organizations, and social media, it is reasonable to forecast a transition from the U.S. being the only superpower to a more multi-polar world. But how that will change global decisionmaking and ethical considerations is not as clear. Although the U.S. has provided some leadership in bringing ethical considerations into many international organizations and forums, its ethical leadership is compromised—there is still no generally accepted way to get corrupting money out of politics and elections or to stop "cozy relationships" between regulators and those they regulate.

Figure 8. Global surface temperature anomalies

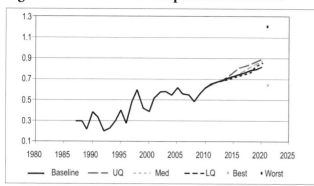

Figure 9. Population growth (annual percent)

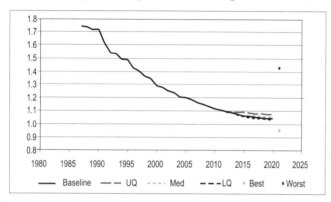

Figure 10. Life expectancy at birth (years)

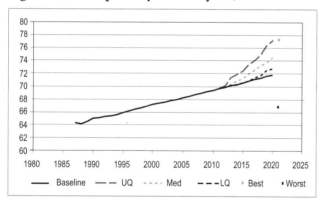

Figure 11. Infant mortality (deaths per 1,000 births)

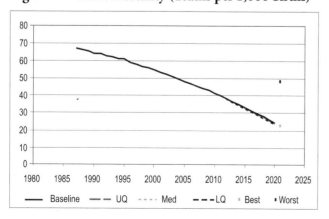

Figure 12. Undernourishment (percent of population)

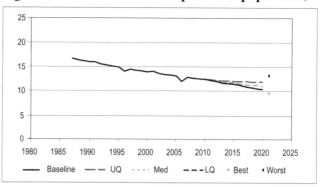

Figure 13. Improved water source (percent of population with access)

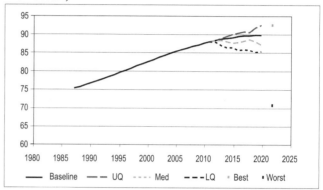

Figure 14. School enrollment, secondary (percent gross)

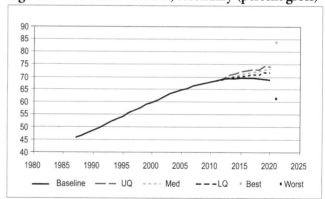

Figure 15. Prevalence of HIV (percent of population of age 15-49)

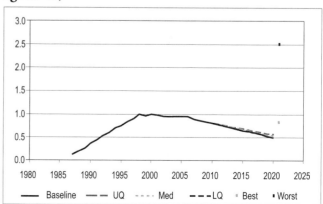

Compilation and projections by The Millennium Project from international data sources listed in Chapter 2. SOFI, on the CD

Figure 16. Women in parliaments (percent of all members)

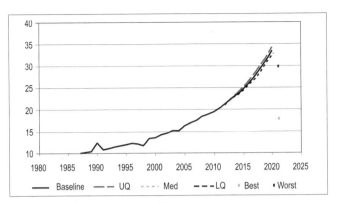

Figure 17. Internet users (billion people)

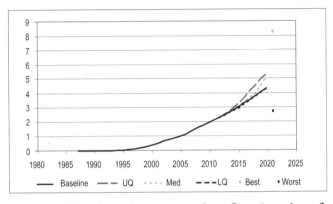

Figure 18. Number of major armed conflicts (number of deaths >1,000)

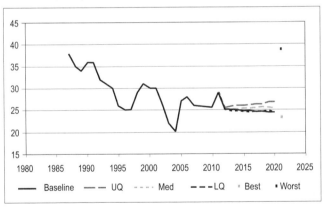

Figure 19. R&D expenditures (percent of national budget)

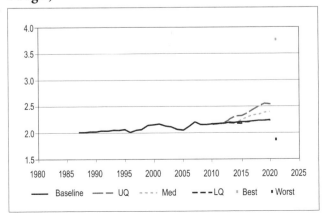

Figure 20. GDP per capita (constant 2000 US$)

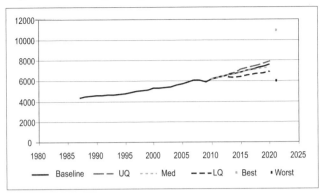

Figure 21. Unemployment, total (percent of total labor force)

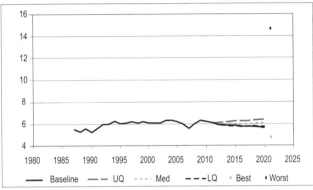

Figure 22. Poverty headcount ratio at $1.25 a day (PPP) (percent of population) (low- and mid-income countries)

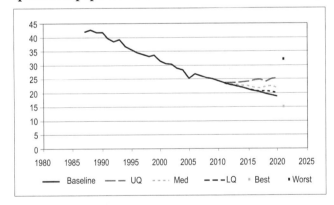

Figure 23. Levels of corruption (15 largest countries) (larger numbers = less corruption)

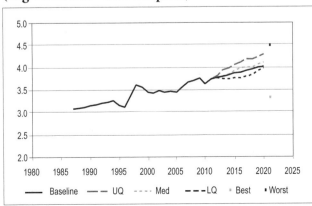

Figure 9 Global Challenges and SOFI Process

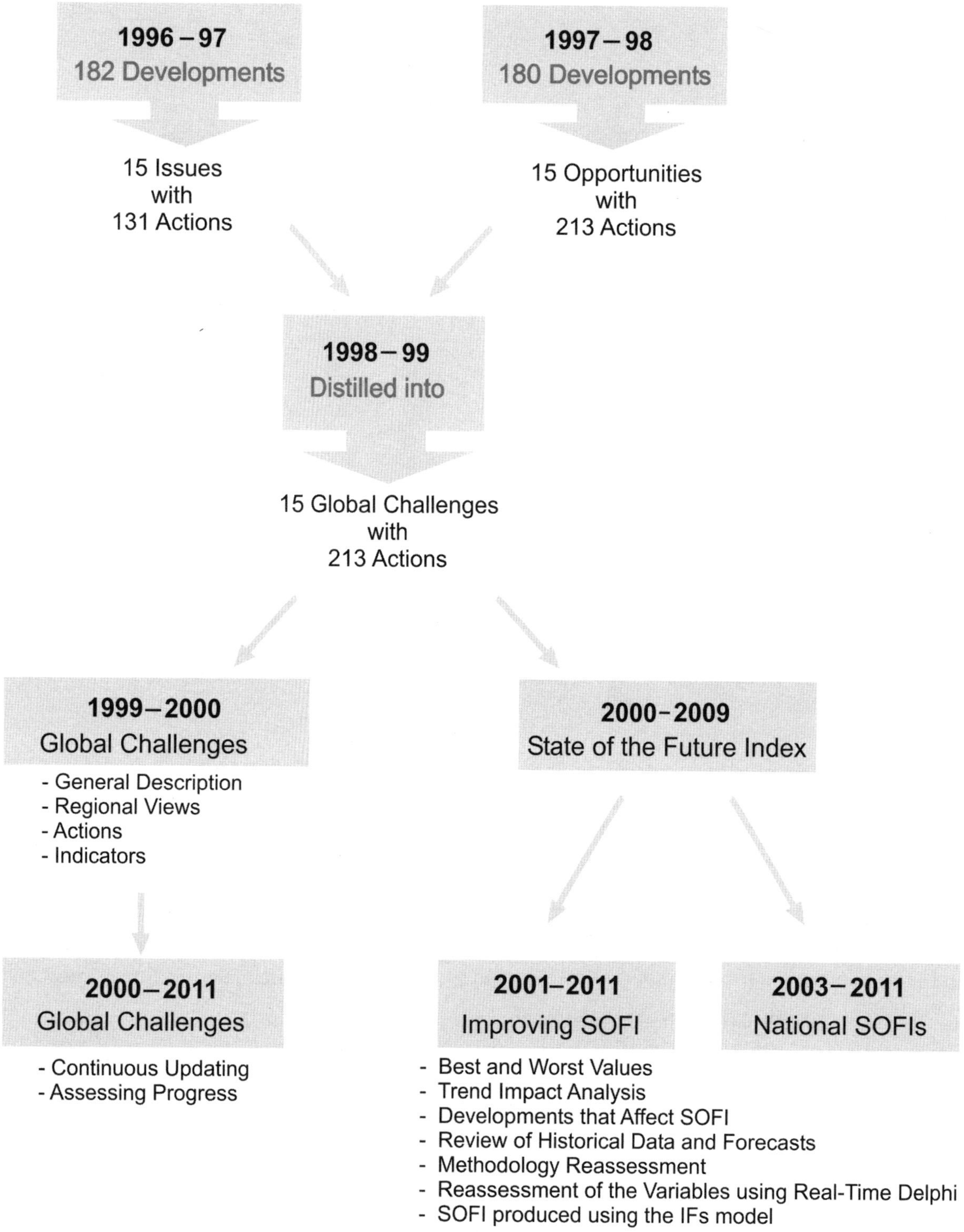

1996 – 97
182 Developments

15 Issues
with
131 Actions

1997 – 98
180 Developments

15 Opportunities
with
213 Actions

1998 – 99
Distilled into

15 Global Challenges
with
213 Actions

1999 – 2000
Global Challenges

- General Description
- Regional Views
- Actions
- Indicators

2000 – 2009
State of the Future Index

2000 – 2011
Global Challenges

- Continuous Updating
- Assessing Progress

2001 – 2011
Improving SOFI

2003 – 2011
National SOFIs

- Best and Worst Values
- Trend Impact Analysis
- Developments that Affect SOFI
- Review of Historical Data and Forecasts
- Methodology Reassessment
- Reassessment of the Variables using Real-Time Delphi
- SOFI produced using the IFs model

2

State of the Future Index: Global Progress and National Applications

In 2001, The Millennium Project began to explore the possibility of creating a quantitative measure, the State of the Future Index, that would depict the global state of the future—measuring, in effect, whether the future seemed to be improving or not over a 10-year period. Inevitably some of the variables included in a SOFI will show the potential for improvement while others show worsening, but the SOFI integrates such changes into a single measure so that the balance between pluses and minuses can be assessed.

The SOFI is unique. Most indexes measure present and past conditions; SOFI provides a projection into the future. Its judgmental aspects are determined by a panel of international experts rather than by a few staff members. It is probabilistic and therefore shows a range rather than a single value of plausible expectations. It is useful in assessing the consequences of contemplated policies. It shows how changes designed to improve one aspect of a complex system can ripple out, affecting others—some favorably and others unfavorably. It shows the net consequences in a way that is easy to understand.

A new global SOFI has been constructed in 2010–11, and a national SOFI has been constructed in Kuwait for the Prime Minister's Office and another in Australia for Timor-Leste to help plan and track Australia's aid to that country.

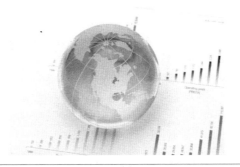

Box 3

The variables included in the 2011 SOFI:

- Improved water source (percent of population with access)
- Literacy rate, adult total (percent of people age 15 and above)
- Levels of corruption (15 largest countries)
- School enrollment, secondary (percent gross)
- Poverty headcount ratio at $1.25 a day (PPP) (percent of population) (low- and mid-income countries)
- Countries having or thought to have plans for nuclear weapons (number)
- Carbon dioxide emissions (global, kt)
- Unemployment, total (percent of total labor force)
- GDP per unit of energy use (constant 2000 PPP $ per kg of oil equivalent)
- Number of major armed conflicts (number of deaths >1,000)
- Population growth (annual percent)
- R&D expenditures (percent of national budget)
- People killed or injured in terrorist attacks (number)
- Non-fossil-fuel consumption (percent of total)
- Undernourishment (percent of population)
- Population in countries that are free (percent of total global population)
- Global surface temperature anomalies
- GDP per capita (constant 2000 US$)
- People voting in elections (percent of population)
- Physicians (per 1,000 people) (surrogate for health care workers)
- Internet users (per 1,000 population)
- Infant mortality (deaths per 1,000 births)
- Forestland (percent of all land area)
- Life expectancy at birth (years)
- Women in parliaments (percent of all members)
- Number of refugees (per 100,000 total population)
- Total debt service (percent of GNI) (low- and mid-income countries)
- Prevalence of HIV (percent of population of age 15–49)

This year's work had several important differences from earlier SOFIs produced by The Millennium Project:

1. SOFI-based projections for scenarios were introduced. By systematically varying the probabilities of the developments included in the analysis, several global projections for scenarios were represented; the SOFIs showed how the state of the future might appear in those futures, providing a quantitative expression.

2. An update of historical data. Changes in data are the result of changes in definitions and data revisions as new information is integrated by the compiling agencies. New series were inserted when old series were discontinued,[1] and new interpolations were made for missing data.

3. Revisions to the developments in the SOFI: reassessed probabilities and impacts. The computation of SOFI involves assessing the consequences of future developments on the forecasts of the variables in a process called trend impact analysis. This and previous work included 90 or so developments. Some were dropped, others were added, and the likelihood and impacts of these developments were changed where necessary.

4. A new statistical program was used to extrapolate the historical data to form the baselines for the SOFI variables.

5. Software was developed to collect group judgments required in assessing the probabilities of the developments and their impacts on the SOFI variables. When used—next year, it is hoped—the software will bring to TIA studies many of the advantages that Real-Time Delphi has provided to Delphi studies: asynchronous participation, worldwide spread, and improved efficiency.

The historical data for each of these variables was fit with time series equations, and the best fit curves that produced plausible future values were used. The statistical software used was CurveExpert Pro (www.curveexpert.net), which attempts to fit more than 50 curve types. The baseline SOFI that resulted from the use of the new data for these variables is shown in Figure 24.

This year's SOFI forecast compares well with the SOFI prepared in earlier years.

In 2010, the University of Denver's International Futures econometric model[2] introduced a SOFI calculation capability. This provides the ability to compute SOFIs for all countries for which data are available, and it suggests the possibility of producing an annual or biennial publication that tracks the SOFI for countries in a manner similar to the annual Human Development Index of the United Nations Development Programme or Transparency International's Corruption Index. As an illustration, the IF modeling system was used to compute the SOFI for six countries for this report, as shown in Figure 25.

Figure 24. 2011 State of the Future Index

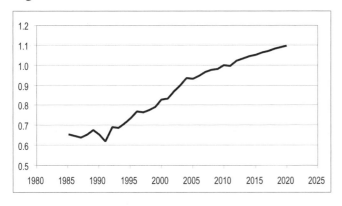

Figure 25. SOFI using IFs for six countries

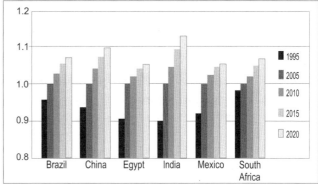

---

[1]   One particular example was the use in 2011 of the variable "undernourishment" which replaced the original series for "food availability (cal/cap)." It became clear in the effort to update food availability that more emphasis today was being placed on the MDG measure undernourishment, a broader measure of world hunger, so the change was made for SOFI.

[2]   Theodore Gordon, Barry Hughes, Jose Solórzano, and Mark Stelzner, "Producing State of the Future Indexes Using the International Futures Model," *Technical Forecasting and Social Change*, January 2011, pp. 75–89.

The Global SOFI

Data Sources and Extrapolation

To simplify the data collection activity, a single source of information was used wherever possible—the World Databank,[3] which provided about 70% of the required data series. The detailed definition of the variables and the equations used to interpolate and extrapolate the data can be found in the CD, Chapter 2.

Calculating the SOFI

This year's calculation is based on the Millennium Project's 2007 RTD produced by global expert panels that provided judgments about the best and worst expectations for the variables in the next 10 years and their weights. These values had to be adjusted to account for changes in definitions and differences between the questions posed in the RTD and the current definitions of the variables. Some of these adjustments involved staff judgments and therefore should be replaced with more recent assessments when the RTD is repeated. Table 1 summarizes the adjusted parameters.

Trend Impact Analysis and SOFI Scenarios

The developments included in the trend impact analysis were based on the list of developments generated in the RTD in 2007. The probabilities were adjusted by staff to account for new events and new perceptions about their likelihood. This set is presented in Table 2 (and in the CD) in the column titled Baseline. To the right of this column are four columns under the heading "Scenarios." These present somewhat different probabilities for the developments under the assumptions of:

- Recession: the world is in for a long tedious recovery, with continuing global recession
- Bad Weather: bad weather and natural disasters prevail
- Bellicose: wars and conflicts expand in severity and frequency
- Green: improvements in the probabilities of promising environmental developments

A table of the sort of Table 2 was prepared for the 93 developments considered; it appears in the CD. The illustration in Table 2 only presents the first 10 developments. The highlighted cells show where changes in the nominal probabilities have been assumed to form the scenarios.

The baseline assumptions were used in TIAs to produce the SOFI forecasts of the sort shown in Figure 26; all of the TIAs are included in the CD.

Changing only the assumptions about the probability of certain developments to simulate the consequences of the global recession and other circumstances, four scenarios were constructed by changing the probabilities of developments included in the TIAs. These simulations were made by a single member of the staff and should therefore be taken with a grain of salt; perhaps in the future such estimates can be provided by groups of experts. Nevertheless this demonstrates the process of using assumptions about event probabilities in a "characteristics matrix" to help assure that each scenario of a set considers all developments that appear in other scenarios in the set and to help achieve some level of completeness and balance. In this demonstration, these assumptions resulted as shown in Figures 27 to 30.

Figure 26. 2011 State of the Future Index with TIA

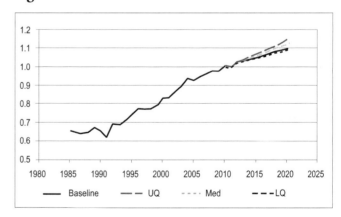

Table 1. SOFI variables with adjusted parameters

Variable	Best 2020	Worst 2020	Weight
Improved water source (percent of population with access)	92.85	78.56	9.08
Literacy rate, adult total (percent of people age 15 and above)	99.51	86.79	8.57
Levels of corruption (15 largest countries)	3.35	4.55	8.54
School enrollment, secondary (percent gross)	83.13	61.96	8.53
Poverty headcount ratio at $1.25 a day (PPP) (percent of population) (low- and mid-income countries)	15.36	32.00	8.57
Countries having or thought to have plans for nuclear weapons (number)	15.29	22.94	8.40
Carbon dioxide emissions (global, kt)	32,209,566	48,774,486	8.40
Unemployment, total (percent of total labor force)	4.92	14.76	8.25
GDP per unit of energy use (constant 2000 PPP $ per kg of oil equivalent)	7.53	6.91	8.02
Number of major armed conflicts (number of deaths >1,000)	23.43	39.05	8.25
Population growth (annual percent)	0.96	1.43	8.19
R&D expenditures (percent of national budget)	3.78	1.89	8.15
People killed or injured in terrorist attacks (number)	25,334.49	956018.34	8.14
Non-fossil-fuel consumption (percent of total)	29.23	19.49	8.05
Undernourishment (percent)	9.71	13.24	7.93
Population in countries that are free (percent of total global population)	59.99	39.99	7.85
Global surface temperature anomalies	0.65	1.22	7.84
GDP per capita (constant 2000 US$)	10,963.12	6029.72	7.59
People voting in elections (percent of population of voting age)	40.95	29.25	6.83
Physicians (per 1,000 people) (surrogate for health care workers)	2.70	1.545	7.67
Internet users (per 1,000 population)	8,310,639.760	2,770,213.253	7.54
Infant mortality (deaths per 1,000 births)	23.11	48.86	7.41
Forestland (percent of all land area)	31.64	24.72	7.43
Life expectancy at birth (years)	77.40	67.08	7.15
Women in parliaments (percent of all members)	29.65	17.79	7.06
Number of refugees (per 100,000 total population)	158.00	531.46	6.97
Total debt service (percent of GNI) (low- and mid-income countries)	4.82	5.52	6.78
Prevalence of HIV (percent of population 15–49)	0.85	2.53	8.19

Table 2. Developments considered in SOFI with their respective probabilities and potential scenarios
(Highlighted cells show where changes in the nominal probabilities have been assumed to form the scenarios.)

Item	Baseline	Scenarios			
		Recession	Bad Weather	Bellicose	Green
The Fukushima nuclear accident causes many nuclear nations to de-nuclearize	50	25	10	50	75
A very good, fast $150 laptop computer becomes available everywhere	85	85	85	85	85
A "teachers without borders" movement develops (50,000 new teachers in the field)	30	30	30	30	30
A pandemic on the scale of HIV/AIDS occurs	30	30	95	30	30
At least 10 countries introduce effective policies designed to increase birth rates to avoid population implosion	75	75	75	75	50
Automation and robotics increase productivity 25% in enough countries to make "jobless" economic growth	50	50	25	50	50
A cheap and effective anti-aging therapy is available	35	35	35	35	35
Bad weather (storms, hurricanes, floods) causes widespread crop failures in at least one year	45	45	100	45	45
Canada begins to export water	35	35	35	35	35
Carbon sequestration is used by 25% of carbon-based industries	50	50	25	50	75

Figure 27. 2011 SOFI, Recession Scenario

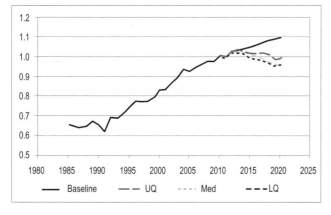

Figure 28. 2011 SOFI, Bad Weather Scenario

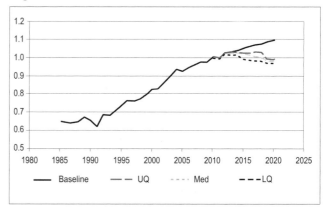

Figure 29. 2011 SOFI, Bellicose Scenario

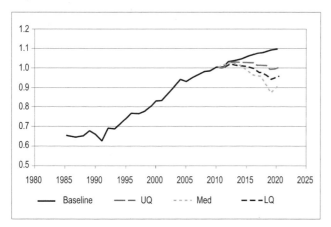

Figure 30. 2011 SOFI, Green Scenario

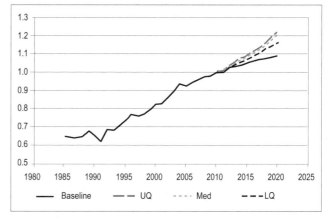

National SOFIs

National SOFIs have been produced for half a dozen countries (South Korea, Timor-Leste, Turkey, South Africa, and others) to assess progress toward an improved future and to test the consequences of proposed policies on a country's outlook. In national SOFIs the variables, weights, future developments, and other judgmental aspects of the analysis are chosen specifically for the country of interest. For example, in some countries the variable "population growth rate" might be viewed as a negative factor, but other countries—concerned about a loss of population—would see an increase in this variable as desirable.

SOFI for the Democratic Republic of Timor-Leste

A SOFI for the Democratic Republic of Timor-Leste was produced in Australia. This work included a TIA and was based on The Millennium Project publications and tutorials.

In Timor-Leste, three of the central issues are:

- population growth (which according to World Bank data hit 4.7% per year in 2004 but dropped to 3.2% in 2009)

- availability of fresh water (World Bank data indicate that improved water sources were available to 63% of the population in 2005 and 69% in 2008)

- malnutrition (over 50% in 2003, the latest World Bank data, as indicated by height of children under 5 years of age).

These and other SOFI components were the subject of TIAs in computing the Timor-Leste SOFI, which is shown in Figure 31. Generally it was found that the developments considered in the TIAs had the consequence of improving the forecasts, although in some instances the spread was large.

Was the exercise helpful? Michael Martin, who is in the Department of International Relations at Flinders University in Australia and the author of the SOFI paper on Timor-Leste, says that the reasons for choosing Timor-Leste for study include: 1) Timor-Leste is similar to other small, poor, slow growth countries in Australia's area of strategic interests; 2) availability of historic data; and 3) Timor-Leste's heavy dependence on foreign aid (Australia being a major supplier of the aid). SOFI served the purpose of monitoring and evaluating needs and effectiveness of Australia's policies toward that country, particularly since the forecasts of the variables that make up SOFI can provide perspective on internal changes and future challenges.

The author of the Timor-Leste report said that the SOFI output shows "Timor's future promises to be marginally better than the present in 2020. The slow, but positive growth is not surprising considering the situation that it currently finds itself in. A rapidly growing population, the result of a high fertility rate is creating a diverse range of problems for Timor including: water supply, food security and basic infrastructure. This information is useful to Australian agencies as it details the drivers of change and allows policy makers to determine where ODA will make the biggest difference in influencing a more positive future in Timor."[4]

Furthermore, in evaluating the SOFI itself, the paper concluded that the method was a very useful policy tool providing a means for them to reach a deeper understanding of the relationships among measures of progress within Timor and the interdependence among these measures. It also illustrates how policies designed to accomplish a primary objective can affect other measures usually unintended and affect the system as a whole, well beyond the intended target. Previous measures of success of particular aid policies have been limited to measures of the projects themselves, while SOFI provides the means to view impacts of such policies on donor and recipient nations as a whole. Their judgment was that SOFI "is a useful tool to provide policy makers with a significant degree of insight into what will drive change within developing nations. It is particularly useful in identifying problematic areas that may arise in the future, thus allowing policy makers to address each of the issues."[5]

Figure 31. SOFI with TIA for the Democratic Republic of Timor-Leste

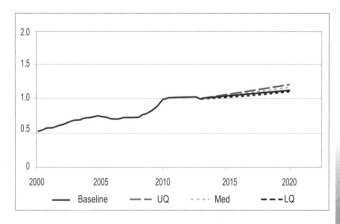

4 See Michael Martin, "The Applicability of a National Focus State of the Future Index (SOFI) on Developing Nations and the Implications for Stabilisation Operations—A Case Study of Timor-Leste," *Futures*, February 2011, pp. 112–19.

5 Ibid.

SOFI for the State of Kuwait [6]

The Kuwait's State of Future Index is prepared to stimulate strategic thinking about Kuwait's future and to highlight key areas for strategy and policy reforms to ensure that Kuwait is on track to realize the objectives and targets set in its long-term vision, Vision Kuwait 2035. The complete Kuwait SOFI report is included in Chapter 2 on the CD.

The K-SOFI was constructed using the results of an RTD survey conducted in July 2010 to obtain expert judgments about variables and developments important to the future of Kuwait. About 50 participants provided numerical answers and narrative comments, for a total of 4,199 answers.

The judgments of the experts together with 20 years of historical data (1990-2009) and forecasts up to 2020 for all variables led to the construction of the K-SOFI baseline. This was followed by a trend impact analysis that considered the consequences of 28 future developments. The 20 years of historical data for each variable were derived from the World Bank database and primary sources, when available. A summary of the historical and forecasted values of selected variables is presented in Table 3 and the resulting K-SOFI with trend impact analysis is in Figure 32.

Overall, the K-SOFI baseline curve shows a steady improvement over the past 20 years (1990-2009). The projection for the subsequent decade (2010-2020) forecasts a continuation of this upward trend, but at a lower rate than the one recorded for the preceding years.

Using 28 developments in the TIA led to the production of forecasts for all variables similar to the one showed in Figure 33. The developments that were included in the TIA and judgments about their probabilities, importance, and the readiness of institutions to deal with them are presented in Table 4.

Major conclusions drawn from this study include:

- Kuwait should diversify its economy away from the oil sector.
- Health sector reforms should top social sector reform priorities.
- Concrete steps are needed to sustain governance reforms and to reverse the deteriorating trend in corruption perceptions.
- Research and development spending should be augmented.
- Recent gains in political liberalization and human rights are encouraging; they should be preserved and sustained.
- Electricity consumption rates are alarming and increasing; measures to reduce electricity consumption per capita should be adopted.

As the inaugural SOFI for Kuwait, this report offers a baseline with which to compare future trends and prospects. Adopting the same method, subsequent indices compare progress against the current standing. Thus, it offers policymakers a recognized benchmark to gauge Kuwait's future path.

Figure 32. The Kuwait State of the Future Index with TIA (2010=1)

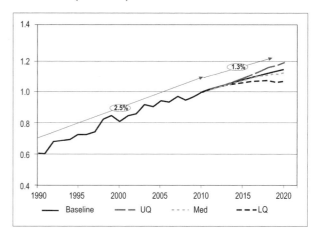

Figure 33. Kuwait CO_2 emissions with TIA (tons/capita/year)

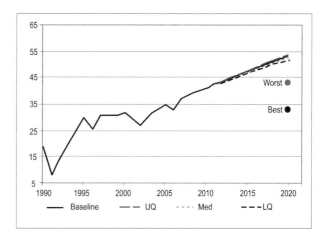

Conclusions

Because of the promise of the work accomplished to date, we believe that SOFI can become a very useful national planning tool. Remaining work includes providing reliable and recent historical data for use in the SOFI calculations, developing a method for automating the trend impact analysis, investigating the use of the Index in sectoral applications (such as for a nano technology or an energy SOFI), and systematically introducing planners to the tool.

---

[6] The material in this section was extracted from "The 2010 Kuwait's State of the Future Index (K-SOFI)," The Early Warning System Strategy Unit, The Technical and Advisory, Diwan of HH The Prime Minister, October 1, 2010.

Table 3. Historical and forecasted values of selected variables used in K-SOFI

Variable	1990	2000	2010	2020
Global demand for oil	65.90	76.20	86.60	98.47
Corruption perception index (Kuwait)	5.73	5.42	4.20	3.73
Kuwait proven oil reserves (billion barrels)	97.03	96.50	103.81	118.40
Global proven oil reserves (billion barrels)	1,003.17	1,104.49	1,278.99	1,443.31
General government final consumption expenditure	39.00	21.00	14.84	8.54
Electricity consumption (KWh/cap/yr, Kuwait)	8,108.00	13,378.00	17,523.62	20,108.57
Gross national income per capita (PPP) Kuwait	24,274.73	35,000.00	57,241.00	70,977.00
Adult literacy rate	77.95	86.89	94.03	94.13
Secondary school enrollments	82.62	93.57	90.82	90.92
Physicians	1.57	1.62	1.88	2.03
Global food availability	2,709.00	2,790.00	2,868.54	2,962.57
Number of major global armed conflicts	35.00	23.00	15.00	11.00
Kuwait CO2 emissions	40,711.00	71,049.00	95,395.60	113,068.60
Military expenditures	48.71	7.15	4.09	2.73
Number of internet users	0.00	150,000.00	1,200,000	2,250,000

Table 4 Developments included in the TIA and judgments about their probabilities, importance, and institutional readiness to deal with them

Developments	Probability (1-100)	Importance (1-10)	Readiness (1-10)
Massive financial crisis triggers a world depression as large as in the 1930s	42.39	7.56	7.14
New technology displaces carbon fuel as cheapest energy source	30.7	7.33	6.43
Gulf Cooperation Council moves countries in the region toward EU-like agreements	46.43	7.22	7.04
Extremist political religious groups change the current direction of the region	47.25	7.21	7.42
Aging population doubles government social costs	65.42	7.16	6.44
Water scarcity problems are essentially solved (e.g., through low-cost desalination)	27.69	7.08	6.77
Renewable energy sources, like wind and solar, provide 50% of the world's power	26.3	6.84	6.70
Electric, hydrogen, and hybrid cars represent 50% of new vehicles sold	32.83	6.18	7.00
Genetic manipulation is used in the production of two-thirds of the world's food	43.98	5.97	6.65
Global protectionist movement dramatically reduces world commerce	22.23	6.02	7.00
Wireless electricity transmission on Earth at gigawatt levels is proved feasible	33.08	6.06	6.53
Extremists detonate nuclear devices, dirty bombs, or other weapons of mass destruction	24.97	7.00	5.92
Iran, Iraq, and Saudi Arabia territorial issues are resolved	21.24	6.00	6.42
Stability achieved in Iraq	45.38	7.00	7.60
Cyber warfare is more difficult to detect and triples in damages from 2010 levels	63.62	6.54	6.45
OPEC's ability to control oil production dramatically dissipates	41.66	6.65	7.29
Middle East oil-producing countries successfully diversify beyond energy production	34.57	6.32	6.17
Human migrations at twice today's levels occur from causes such as water shortages	45.27	5.86	6.05
Carbon sequestration: 25% of all new carbon production is captured and stored	33.81	4.20	6.08
Energy companies are found responsible and sued for previous CO_2 emissions	18.52	4.81	5.32
Conservation efforts throughout the world reduce energy demand by 10%	41.94	6.26	6.52
E-Learning reaches tipping point, increasing access to high-quality education	56.07	5.85	6.36
Economic growth spike in other global regions limits ability to attract talent	73.46	6.44	6.79
Gender parity in employment is achieved	46.62	6.14	6.79
Artificial bacteria and algae are used to produce 20% of fuels	23.52	5.26	6.09
Most glaciers melt twice as fast as in the decade 2000 – 09	51.19	6.67	6.21
Global pandemic kills over 100 million people	18.22	4.75	6.86
Megawatt levels of power are beamed from solar collectors in orbit to Earth	18.95	4.47	4.71

3

Egypt 2020

The world cheered the Egyptian Revolution and now wonders, What's next? Will Egypt invent the first twenty-first century new form of democracy, taking into account the role of cyberspace, international interdependency, and a rapidly changing world? Will it become a centrally controlled political system with a decentralized local economic system? Or will it create a participatory democracy using the power of the Internet to constantly identify new approaches through a national collective intelligence system to address the persistent problems of poverty, water, education, and public health?

The Cairo Node of The Millennium Project created a Real-Time Delphi on the future of Egypt. The results will be used by the Egypt Arab Futures Research Association and its Collaborative Partners to create future scenarios of Egypt and produce a *State of the Future of Egypt* report. The study is ongoing at the time of this writing; hence, this chapter is a work in process.

The study was designed to collect judgments about developments that might shape the future of Egypt, exploring new directions for the country after the revolution. These developments or seeds of the future were identified by an initial group of Egyptian futurists, historians, sociologists, and professionals from a variety of fields. The initial group submitted suggestions about emerging trends, technologies, and priorities to improve the future for all Egyptians, along with revolutionary genres and formats.

The Cairo Node invited selected Egyptians and those who have worked in Egypt to rate 34 developments as to how strongly the public supports the development; the level of government support for the development; how likely it is the development will be achieved by 2020; and how to ensure the continued expansion of the new Egyptian spirit and an Egyptian renaissance of thinking for a more positive future. Participants were also invited to explain their answers without attribution. The methodological background is fully explained in the CD section Egypt 2020. Table 5 lists the 34 developments in the order of how likely it is the development will be achieved by 2020.

What will ensure the continued expansion of the new Egyptian spirit and an Egyptian renaissance of thinking for a more positive future? A distillation of views

The continued and persistent uprising of young people (both nationally and internationally)… making real gains politically, socially, economically, and spiritually… People will insist on not missing the chance to create a bright future… hand in hand charting the future now and only now.

People are no longer afraid to protest; they know how to defend their rights. They know how to demonstrate peacefully in "Friday's Million Marches" and how to confront security forces of despotic regimes.

Participation of the public and private sectors in adopting the values of the Egyptian Revolution, to support social solidarity and help the needy… Change the tactics of the revolutionary movement and discourse, to identify and present the key issues for the future of Egypt to the people in all the governorates and villages across the country… Establish one or two parties for the civic state.

Procedures for participation on all levels have to be established. A fair and balanced division and sharing of power and the control of power has to be institutionalized. The old power structures have to be eliminated or at least weakened. Fast and sustainable successes in socioeconomic development are necessary.

Tough and drastic measures should be taken within the framework of the law to fend off the sectarian clashes and cool down religious tensions…Amending much of the legislation to protect the public funds against the malpractice and misconduct of officials.

Table 5. Developments that might shape the future of Egypt, listed by their average likelihood to be achieved by 2020

Rank	Item	Popular Support	Gov. Support	Likelihood
1	High-speed Internet access for at least 75% of population	9.60	83.89	92.22
2	Freedom for all to establish political parties	9.10	83.56	90.10
3	Functional illiteracy rate reduced by 50%	8.67	95.75	88.50
4	Sufficient safe drinking water for all	8.20	98.75	86.89
5	Freedom: discussions of issues of tolerance and values exist in all media (TV, radio, press, Internet)	9.07	80.85	85.33
6	Sufficient food for all, with adequate reserves	7.75	96.71	84.86
7	Standard of living for all citizens increases 50%	7.44	92.25	84.71
8	Over 50% reduction in suicide rate among young people	8.50	82.00	84.44
9	Public option poll finds over 50% of the public believes a renaissance has begun in Egypt	9.30	82.88	84.25
10	Rich-poor gap reduced by 50%	8.22	95.00	83.88
11	Equal pay for equal jobs between men and women	8.78	76.50	83.75
12	Establishment of win-win relation between Egypt and Nile-Basin countries	8.90	100.00	83.56
13	Free and transparent voting in election campaigns	8.78	96.50	83.10
14	Very active tele-Egypt connecting Egyptians overseas to development process back home	8.10	90.67	82.22
15	Acknowledgment by most of a new sense of citizenry and national loyalty, particularly among young people	9.20	88.67	82.10
16	Freedom House (an organization that rates countries' freedom) has changed Egypt's status from "not free" to "free"	9.60	92.22	81.27
17	Basic health services accessible to all	7.30	93.75	81.00
18	Political violence essentially ended	7.56	88.63	80.11
19	No sectarian and minority violent incidents reported for over six months	7.10	99.25	79.75
20	Public option poll finds over 50% of the public are happy with the new political and social regimes	9.15	86.71	79.65
21	Corruption reduced by 50%	7.20	94.22	78.50
22	Sanitary sewage access for all	7.67	90.63	78.13
23	Social class and values conflicts are essentially abolished	7.50	79.67	76.00
24	NGOs certify respect for citizens' basic rights by international standards	8.23	83.30	74.82
25	WHO has certified continued progress in all basic health standards	7.89	93.75	74.75
26	Educational objective to increase students' intelligence (brain functioning)	7.44	95.25	74.38
27	Illegal immigration rate reduced by 50%	7.70	80.38	74.00
28	Resolution of conflicts over modernism vs. fundamentalism	7.10	83.44	72.78
29	Air, water, and land pollution decreased by 30%	8.00	86.00	71.75
30	Renewable energy provides 20% of the electricity generation mix	7.70	91.11	71.63
31	Micro-finance and small business development accessible to all	7.60	75.89	67.67
32	Noticeable reduction in business influence in political and economic life	7.50	86.00	65.60
33	Population growth rate reduced below 1%	8.40	95.44	65.00
34	At least 30% of Parliament and Cabinet are women	7.30	64.78	64.44

CHAPTER THREE

What could counter the new Egyptian spirit? A distillation of views

Re-emergence of corruption among political parties dominated by the former politicians of the former regime – Unexplained delays, hesitancy, and/or reluctance in dealing with symbols of the former corrupt regime and/or a postponement of the reforms could counter the new Egyptian spirit. Expedited purging of the ousted autocratic and despotic regime's associates must be implemented as soon as possible.

A state of unrest created by uncontrollable reactions of some people due to ignorance, or thugs leading to unsafe conditions forcing the Supreme Military Council to use violence against destructive protesters – The sudden disappearance of the police created chaos all over the country, and remnants of the toppled regime still cause insecurity. Some speak of the urgent need to restore security and stability, while blaming revolutionaries. Prisoners fled, thugs were freed, and extremists became bolder in public. The Supreme Council of the Armed Forces has to restore internal security. Unlawful acts and violence by the deviants will be met by lawful measures and even violence by the SCAF, if law and legality are to reappear.

External turmoil from neighboring countries – Egypt is committed to observing the human rights of Egyptians and its neighbors. Libyans and Tunisians fled to Egypt. Palestinians in Gaza were seeking relaxation of their long-lived siege. Revolutionaries in Syria, Yemen, Bahrain, and other Arab countries are inspired by the Egyptian revolt.

External interference of some countries to achieve certain objectives where Egypt's political disorder is in their favor – Some of the neighboring influential countries are pressing for reconciliation with the ousted president and his associates, despite their slaughtering, profiteering, and corrupt activities. External foreign concepts are addressed in favor of new region democratization, a greater Middle East, creative chaos, and combating terrorism.

Lack of commitment by the government pledges – There were rumors that the government and SCAF might forgive the ousted president and his associates, despite the crimes committed during the last 30 years. To help counter this, a mass march was held at Al-Tahrir Square on May 27, 2011, as the "Second Anger Friday" and the "Second Revolution."

Divisions within and among the revolutionaries – Divisions among the revolutionaries, especially between the youth coalition and the Muslim Brotherhood and other political powers, could create a rift between the people and the ruling military council (SCAF). Calls to violate SCAF's roadmap (which was approved by the majority, including the proposed constitutional amendments, during the March 19 referendum) would counter the spirit that is uniting the masses and the ruling military council.

Resorting to foreign aid with tough preconditions to avoid national bankruptcy – SCAF warned that the Egyptian economy could collapse, leading to capital flight and the use of its strategic reserves. If this begins, then the government would agree to foreign aid from international institutions or the big powers, which carries tough preconditions, which might constitute an additional burden to the economy rather than salvaging it and might depress the Egyptian spirit.

The Egypt 2020 study is still ongoing as this report is being printed. Those interested in the study should contact Dr. Kamal Zaki Mahmoud Shaeer, Chair of The Millennium Project Node in Cairo and the Egyptian-Arab Futures Research Association, at kzmahmoud@hotmail.com.

4

Future Arts, Media, and Entertainment: Seeds for 2020

The explosive, accelerating growth of knowledge in a rapidly changing and increasingly interdependent world gives us so much to know about so many things that it seems impossible to keep up. At the same time, we are flooded with so much trivial news that serious attention to serious issues gets little interest, and too much time is wasted going through useless information. How can we learn what is important to know to make sure that there is a good future for civilization? Traditionally, the world has gained access to most of its knowledge through the education systems, the arts, media, and entertainment. Today and into the future, information flow will be even more pervasive, with ever increasing communication technologies emerging on the landscape of experience. In 2007 The Millennium Project conducted a global assessment of some elements of the future of education and learning (see CD Chapter 10 Future Possibilities for Education and Learning by the Year 2030). This year The Millennium Project looked at the future of the arts, media, and entertainment. A distillation of the results is presented in this chapter, while the details are available in Chapter 13 of the CD.

Clearly the arts, media, and entertainment have an enormous influence on people's worldview and, as a result, on the future. Every day the average global citizen watches 3.4 hours of television, listens to 2.2 hours of radio, uses the Internet 1.7 hours, reads magazines and newspapers for 1 hour, and watches 8.5 minutes of cinema.[7] Our homes and workplaces are filled with art and various forms of media. And with the wireless revolution and smart phones, media is an ever-present reality for an increasing number of citizens. Not only can these powerful tools of transformation inform and influence humanity's understanding of itself, they can also aid in the evolution of society by inspiring visions, disseminating information, and catalyzing actions that address the 15 Global Challenges in Chapter 1 and other areas in need of change.

With thousands of global channels to choose from plus the Internet, social media, mobile phones, computer tablets, games, and a proliferation of new media technologies, it is essential that we become media literate and also create from a new socially beneficial awareness. While there has been tremendously powerful media created in the past, there is so much potential that we have yet to tap through the intentional use of these powerful creative tools to educate and address pressing issues facing humanity. The 15 Global Challenges and the solutions offered in the *2011 State of the Future* can inspire and inform artists, content creators, and those working in these fields to create art, content, and stories that transform when distributed to the masses.

---

[7] *TGI Global Update*, Issue 3, Summer 2009 (http://globaltgi.com/knowledgehub/documents/TGIUpdate3.pdf)

Inspired by the Florentine Camerata Society, a sixteenth-century "think tank" responsible for the creation of the art form we know today as the European opera, The Millennium Project created a new Node, a Global Arts and Media Node. Under the leadership of arts and media professionals and educators dedicated to a more positive future, this Node is responsible for tracking and reporting on the latest trends in the arts and media as well as serving as a resource for artists, content creators, and media makers interested in creatively advocating solutions to the 15 Global Challenges found in this report.

Futuristic innovators around the world were invited to suggest and discuss future elements or seeds of the future in the arts, media, and entertainment. After a month of online discussions, 34 elements were chosen and put into a Real-Time Delphi for an online international assessment. Several of the seeds chosen are new trends and technologies that are emerging in the media and entertainment landscape. Other seeds are topical and related to the use of the arts, media, and communications technologies. Writers, producers, performing artists, arts/media educators, and other professionals in entertainment, gaming, and communications were nominated by the 40 Millennium Project Nodes around the world to share their views. They were asked how likely it was that each element or seed might become dominate by 2020. They were also asked how important these seeds are for achieving the best for civilization and if they were or would be interested in developing the seed with other elements of the future. About 250 people from 33 countries signed on to the study. Of these, over 150 provided at least one response. This 60% response rate is fairly representative of similar studies.

Figure 34: An example of the questionnaire for 2 of the 32 elements:

Questionnaire

	Questions	Likelihood	Importance	Comments
1	Cyber-Techno Classical (Blending Classical Arts and Futuristic Technologies) *click here for example* *click here for example*	Please enter the probability that this seed of the future will grow into a dominant form in popular media by 2020 (100=certain) 95 Changes are OK Average= 64.9 (141) Submit only this cell (go) Reasons for your answer .. *click here*	How important will this be in achieving the best for civilization? (100 = hugely; 1 = not at all). 80 Changes are OK Average= 48.3 (130) Submit only this cell (go) Reasons for your answer .. *click here*	Are you or would you like to be involved in developing this seed with other futuristic seeds? Please explain. To proceed to the answer form please *click here*
2	"Holographic" Performances (live performance of virtual characters - digital puppetry) *click here for example* *click here for example*	Please enter the probability that this seed of the future will grow into a dominant form in popular media by 2020 (100=certain) Average= 58.8 (119) Submit only this cell (go) Reasons for your answer .. *click here*	How important will this be in achieving the best for civilization? (100 = hugely; 1 = not at all). Average= 43.3 (112) Submit only this cell (go) Reasons for your answer .. *click here*	Are you or would you like to be involved in developing this seed with other futuristic seeds? Please explain. To proceed to the answer form please *click here*

The results are displayed on the following four pages, listed in order of likelihood of becoming a dominant form by 2020.

Arts/Media results ordered by likelihood of becoming dominant by 2020

1. Multi-touch Displays

Likelihood: 91.18, **Importance:** 63.54
Example: http://www.ted.com/talks/jeff_han_demos_his_breakthrough_touchscreen.html

2. Electronic Publishing (Vooks: integrating text, video, Internet)

Likelihood: 89.09, **Importance:** 76.92
Example: http://shelf-life.ew.com/2009/10/01/what-is-a-vook-and-will-it-change-how-you-read

3. Augmented Reality: (Overlays and Geotagging)

Likelihood: 87.82, **Importance:** 68.85
Example: http://www.mirror.co.uk/news/technology/2009/12/02/twitter-360-geo-tagging-augmented-reality-app-released-115875-21868106/

4. Geographical Information Systems (GIS) - Advanced geographic mapping, visualization and augmentation.

Likelihood: 87.80, **Importance:** 72.45
Example: http://www.ted.com/talks/blaise_aguera.html

5. Convergence of Computer/Mobi Content on Television

Likelihood: 86.39, **Importance:** 60.54
Example: http://www.youtube.com/watch?v=14hnh8yy_H4

6. Ubiquitous Computing

Likelihood: 85.70, **Importance:** 60.06
Example: http://www.youtube.com/watch?v=2I3T_kLCBAw

7. Digital/Social Networking for Cultural Diplomacy & Change

Likelihood: 85.31 **Importance:** 82.68
Example: http://gephi.org/2011/the-egyptian-revolution-on-twitter/

8. User Generated Content Technologies providing Democratization of Content Creation

Likelihood: 82.58 **Importance:** 74.88
Example: http://www.lulu.com

9. Telepresence

Likelihood: 81.49 **Importance:** 73.59
Example: http://www.youtube.com/watch?v=rcfNC_x0VvE

10. Social/Global advocacy through storytelling

Likelihood: 78.65 **Importance:** 73.77
Example: http://www.grist.org/article/2011-01-27-how-to-get-tv-shows-to-tell-truth-about-climate-change

11. Interactive Displays (gesture-based user interfaces)

Likelihood: 78.35 **Importance:** 56.70
Example: http://www.youtube.com/watch?v=GmqJr2ijKIo

12. Media/Arts for Cultural Diplomacy and Change

Likelihood: 77.17 **Importance:** 75.52
Example: http://www.socialchangefilmfestival.org

13. Virtual & 3D Art Exhibits/Collections/Museums

Likelihood: 75.78 **Importance:** 56.28
Example: http://www.googleartproject.com

14. Photogrammetry and Gigapixel Panoramic Imaging; Extracting 3D models or ultra-high-resolution images of spaces

Likelihood: 75.08 **Importance:** 56.08
Example: http://www.photogrammetry.com/

15. Media literacy

Likelihood: 75.03 **Importance:** 76.41
Example: http://www.medialiteracy.com/

16. Multiplayer Online Virtual Worlds

Likelihood: 73.03 **Importance:** 48.67
Example: http://www.virtualworldsreview.com/info/categories.shtml

17. Autostereo (glasses-free) 3D Displays

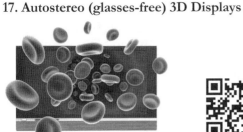

Likelihood: 72.79 **Importance:** 44.30
Example: http://www.magnetic3d.com/

18. Serious Games (Cyber games to educate & solve world problems)

Likelihood: 72.50 **Importance:** 74.39
Example: http://www.urgentevoke.com

19. Augmented Reality: Third-Person Augmented Reality

Likelihood: 68.95 **Importance:** 51.47

Example: http://www.iphoneness.com/iphone-apps/best-augmented-reality-iphone-applications/

20. SciArt (art inspired by science)

Likelihood: 64.81 **Importance:** 58.52
Example: http://artsci.ucla.edu/

21. Cyber-Techno Classical (Blending Classical Arts & Futuristic Technologies)

Likelihood: 64.49 **Importance:** 47.97
Example: http://www.youtube.com/watch?v=D7o7BrlbaDs

22. User-Based Content Creation (i.e. cinematography) within Virtual Worlds (called Machinima)

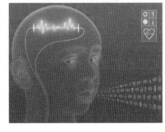

Likelihood: 64.29 **Importance:** 50.36
Example: http://www.machinima.com/

23. Media/Arts/Cyber Healing

Likelihood: 63.85 **Importance:** 62.45
Example: http://www.uclartsandhealing.net/background.aspx

24. Location-based group interactive video games

Likelihood: 63.64 **Importance:** 47.42
Example: http://www.youtube.com/watch?v=y6izXII54Qc

25. Domes, Planetariums and 360 Immersive Cinema

Likelihood: 61.79 **Importance:** 48.39
Example: http://vimeo.com/6988758

26. Augmented Reality: Architectural Projection Mapping

Likelihood: 60.82 **Importance:** 49.74
Example: http://www.fuelyourcreativity.com/architectural-projection-mapping-the-future-of-motion-graphics/

27. Performance in Public Spaces/ Flash Mob Art

Likelihood: 60.42 **Importance:** 52.52
Example: http://www.youtube.com/watch?v=SXh7JR9oKVE

28. "Holographic" Performances (digital puppetry)

Likelihood: 58.34 **Importance:** 42.45
Example: http://www.youtube.com/watch?v=G-AhYnjKEzs

29. ARG: Alternative Reality Games and Experiences

Likelihood: 57.39 **Importance:** 44.05
Example: http://www.youtube.com/watch?v=7iti8Ivy--s

30. Kinetic Art

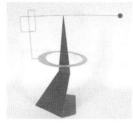

Likelihood: 54.93 **Importance:** 37.64
Example: http://www.kinetica-museum.org/new_site/home.php

31. Technoetic arts

Likelihood: 52.93 **Importance:** 47.02
Example: http://wn.com/Technoetic

32. Eyeware and Head-Mounted Displays (HMD)

Likelihood: 52.82 **Importance:** 37.94
Example: http://www.sensics.com/

Domes, Planetariums and 360 Immersive Cinema: courtesy of Denver Museum of Nature and Science
Kinetic Art: courtesy of Bruce Gray, www.brucegray.com

Table 6. Top Six Seeds in Order of Importance

Seed Topic	Importance		Likelihood	
	Rank	Average	Rank	Average
Digital/Social Networking for Cultural Diplomacy and Change	1	82.69	7	85.31
Electronic Publishing (Vooks or video books)	2	76.93	2	89.09
Media literacy (consumer awareness regarding media creation, content, language, business, etc.)	3	76.41	15	75.03
Media/Arts for Cultural Diplomacy and Change	4	75.52	12	77.17
User Generated Content Technologies providing Democratization of Content Creation	5	74.88	8	82.58
Serious Games (Cyber games to educate and solve world problems)	6	74.39	17	72.5

When analyzing the results of the seeds in terms of "how important these seeds are to humanity" versus the "likelihood that they would happen," it is critical to note that of the top six seeds participants felt were important, only one seed was in the top 10 of those most likely to become popularized. While the other five were considered important for the future, they were also considered less likely to become popular. The study and subsequent studies like it are important in that they help us identify areas of the arts and media that are deemed important to humanity but are in greater need of attention, visibility, or support. These seeds include:

- Digital/Social Networking for Cultural Diplomacy and Change
- Media Literacy
- Media/Arts for Cultural Diplomacy and Change
- User Generated Content Technologies providing Democratization of Content Creation
- Serious Games (cyber games to educate and solve world problems).

Of these, Media/Arts for Cultural Diplomacy and Change, Media Literacy, and Serious Games received the lowest likelihood ratings, indicating that they are fertile areas most worthy of support.

In reviewing the top six seeds in order of importance, "social networking for cultural change" has by far the highest ranking, indicating a need to more thoroughly evaluate this methodology for its ability to influence future trends. Also of high value is the interest in more-accessible digital content and content that is user-generated. Media literacy places high enough in the rankings that consideration should be given to enacting strong evaluations of how we are teaching it both in the classroom and to adults. Serious gaming and socially beneficial storytelling/content creation are both at high importance value as well, indicating that intentional use of these technologies is likely to yield a greater degree of societal change. Societal change can be maximized by a concerted effort to strategically design media and the arts to be effective tools of personal, social, and global transformation.

Narrative Inputs

One distillation of the many views by the participations about the future of arts, media, and entertainment is that it will be global, participatory, tele-present, holographic, augmented reality conducted on next-generation mobile smart phones and immersive screens that engage new audiences in the ways they prefer to be reached and involved.

For each of the 32 seeds, respondents were asked to provide reasons for their quantitative responses and to describe the extent of their current or future involvement in the various seeds/items presented in the questionnaire.

Nearly a hundred pages of comments were received. The complete text of all the responses appears in Chapter 13 of the CD. This section provides a distilled flavor of them in italics:

"Merging of technology and art appears a given. The quality will be in proportion to the genius applied.... I already compose with collaborators via the Internet and want to be part of a virtual choir like Aurumque! It makes room for the spiritual/energy connection between participators, since there is no physical contact to color the relationships. Of course singing in a choir or making music in person with others is the ultimate oxytocin rush and very satisfying as an artist, but virtual performances and concerts are definitely exciting and I can't wait to be involved in one.

Not only will use of futuristic technologies keep the art forms relevant, they will add a fresh experience to the existing body of work created by artists. With older audiences dying out, creating experiences that will attract newer, younger audiences is important to keep creative companies thriving. And these new technologies can also make it possible for us to collaborate and create with artists globally. Use of holographic characters and performers may be a novelty and passing fad, but imagine if you could teleport a real-time performer, speaker, singer, dancer, and so on into a performance space in real-time?

We should develop this flash mob further for "achieving the best for civilization" because these flash mobs bring people together and create a sense of well-being, a sense of happiness. One could argue that it is a shallow sense, however in these hard times where disasters and inhumanity are frequent and will probably become more frequent, I am convinced that these quick, small gatherings where everyone is connected by movement, across language borders and across cultural borders, even for 10 seconds are of essential meaning for our future.

Electronic Publishing is fabulous and definitely here as the new way to read and interact in different mediums. I love the way you can link to others while reading and watching videos of scenes etc. This was all science fiction not long ago, and we've finally brought it into being. The only problem with electronics is that there is more unrecyclable waste, even though it cuts down on tree usage…we are now poised to see a second wave of enlightenment inspired by the twenty-first-century version of the Guttenberg Press, the Internet and e-publishing…Achieving the best for civilization when it comes to electronic publishing I feel is a two-edged sword. As long as the current copyright regulations are in place, the west is well served with all sorts of knowledge. But how about access to knowledge in other countries? This is already a problem now, because e.g. libraries in third world countries have a hard time paying the licenses. If this is not improved, the electronic publishing will go ahead but it will not be achieving the best for civilization worldwide!

Games could assist in sensitizing people to alternative realities and ways of working towards those realities… So much of game-play is based on eliminating "the other;" to solve world problems, we will need to change that view into "each other as our self." …Quite likely in the not so distant future a game developer will win the Nobel Prize for Peace…

Advances in neuroscience will allow a growing understanding about triggering specific mind states and hence develop the field of media/arts/cyber healing. This is already being used for neurofeedback devices to teach people how to entrain specific brain frequencies and has recently broken into the consumer market with the Jedi mind games for kids…Arts-based healing equals deeper-faster compared to traditional talk-based therapy (both are needed, however).

Skype can already allow a certain amount of "telepresence" at little or no cost. By 2020, driven by the costs of travel and the rapid deployment of new tools and capabilities in networking and display systems, this will be highly developed. Some offices and homes in 2020 will have rooms in which large-screen projectors line the walls and create a "holodeck" feel ala "Star Trek" (not that complete immersion, but a representation of it) and you can "transport" yourself to Hawaii or talk to grandma 1,000 miles away, visiting as if she's in the same room with you. We have the technology now to do this, but it's too expensive for consumer adoption…this form will be critical for building more peaceful relations among peoples.

Not only does telepresence technology provide businesses and organizations the ability to meet virtually, it is a technology that artists and creators can explore for group collaboration. Imagine attending a global orchestral performance of the greatest musicians from around the world performing virtually? The World Opera and YouTube Orchestra have been doing work that leads to greater experiments in this effort. This also provides a great opportunity for experiences that encourage cultural diplomacy.

The primary breakthrough in new eyeware and head-mounted displays will be contact lenses with nanobots creating overlay. Current HMD are still clunky and only appropriate for specialty use. Such contact lenses have already been developed in first gen experiments.

"Dumbing down" the user interface makes it less useful to power users who are most likely to be able to innovate and invent new things if they continue to have the most latitude possible. While there's no doubt that multi-touch displays have an advantage because they are intuitive, they can never possibly gain the nuances of control we already experience through current methods of accessing computing power in our large and small devices. Apple and other companies are using software and hardware design to move consumers into a realm in which they will be app-dependent, and while some applications will be "free," most will cost people something every time they choose to use them. The world is moving from freely available information on computers hooked up to information networks to a world of controlled appliances for which they must pay and pay and pay and pay. Apple's latest OS update, nicknamed Lion, is the company's move to get consumers into that controlled world and build up the company's profit-margin. This does create new jobs for applications developers, and monetizing the Internet in more direct ways creates more revenue flows, but the act of monetization politicizes the act of creation and drives people to impose more rules and build more bottlenecks and can cut the poor and uneducated out of the invention and innovation equation.

Many of these seeds of the future raise the question about what is really true. Will we be able to identify true history from reconstruction? When there are competing histories, who will be able to tell which is authentic?

We are seeing the collapse of the distribution system, allowing artists to enter the field as they will. Copyright issues notwithstanding, we are seeing a democratization of art.

Storytelling is the most powerful way to elevate human consciousness. If the story is engaging, people remember it forever and are changed as a result. We have scientific measures in published literature of the profound impact of TV health storylines on viewers' knowledge, attitudes, and behavior. These stories reach up to 20 million viewers in a single hour and up to 80 million viewers in the first week. The episodes later reach up to 400 million viewers in over 100 countries worldwide. When the content has social value, it is a service to humanity.

The way to partner with scriptwriters and producers is NOT through advocacy or "product placement." That is an old paradigm, and Hollywood's writers are allergic to it. I would change the wording of this question to "Social/Global enlightenment through storytelling (partnering with writers/producers to inspire and inform them to address topics of social concern accurately) and to link traditional media and new media in innovative and interactive ways."

In the realm of this Delphi, augmented reality will have quite a high impact compared with other choices. Just-in-time or real-time information can be life-saving in some situations, and it will certainly make a huge difference in informing people, providing them with needed economic, social, and political information on which to base decisions as they move from moment to moment in their daily lives.

The final column asked the question "Are you or would you like to be involved in developing this seed with other futuristic seeds?" The full text of the responses without attribution is in the Appendix of the CD. Here is a short distillation:

I am extremely interested in facilitating the merger of cyber-techno and classical arts; I believe the blend serves as an essential "bridge" between the real world and the cyber world; a bridge necessary to our continual evolution… definitely would like to be involved, as a director, in creating holographic experiences coupled with movion-based (Swedish equal shares), group interactivity with the projections…Being a composer, there are already a couple

of songs I'd love to use in a flash mob. When workshops begin for the new musical I'm writing, which is about tapping into the abundant, positive, and powerful energy available to all humans, I'll be looking for volunteers. … Would like to participate in an event, especially in at-risk urban communities and in senior centers…Would love to participate in developing, participating, or tracking these flash mob seeds because I believe that anything we can do to foster bringing people together is important for our future… there's an obvious link to social media.

Aesthetic experience of global interdependency and harmony requires telepresence — happy to help get the arts more involved in this medium for simultaneous performances on a global basis… ubiquitous computing will be the tech support for the evolution of global consciousness. How well this evolution works will be dependent on how well mystics and technocrats cooperate to make it enlightening… I am leading this work now and urge the UN and others to partner with us. It is critical to separate advocacy of a particular governmental agency in the media from accurate portrayal in entertainment media of some of the most important issues of our time: health, climate change/environment, human rights, humanitarian aid, global leadership, peace/conflict transformation, spirituality, freedom of creative expression, global monetary systems.

Storytelling is the oldest form of education known to humanity, we must use it to tell the crucial stories of our time… certainly climate change is one of those. My research in the area of "evolutionary guidance media" suggests the importance of including data across 10 dimensions of human activity in our story & media creation if we are to more quickly promote planetary consciousness… Yes — this is one of my ultimate goals, to create and direct entertaining social/global advocacy and education media utilizing a number of the Delphi-listed media technologies in convergence.

The participants were also asked if there were any additional media/arts technologies, genres, or modalities that they feel are important that we should be tracking.

Transmedia — the combining of multiple creative practices on a diverse range of platforms for an interactive experience — is increasing …Babel Fish Smart phones… Perhaps the biggest transformation of the future will be started from the demise of the language barrier.

These media technologies have two important attributes: they make possible the construction of realistic alternate histories and others enhance the ability to

persuade. Both are potentially dangerous. They lead to confusion about what is or was really true.

When artificial constructs are very good and indistinguishable from reality there will be two possible disasters. First, the constructs may be seen as good or better than the true world, and people, including poor people, may choose to live there and avoid reality. This happens now with Second Life. Could this become even more addictive when jobs are scarce and free time weighs heavily? Second, perhaps even more disturbing is the possibility of using these media technologies to construct artificial histories that are indistinguishable from real truths. Imagine the holocaust deniers building a Nazi history which omits the holocaust, or Apollo deniers building a site that shows how the lunar landing was faked in Utah: a fake of a fake.

In this era, who could say what was or is really true? Maybe there could be an incorruptible NGO that acts as a global authenticator.

Each of these technologies is fascinating in its own right, and each has the potential for beneficial impact. I believe what will be most interesting are the "mash-ups"—the convergence of a number of these technologies into singular media events/environments. Herein lies my primary focus – both as a forecaster and, mostly, as a creator.

And last, participants were asked if they would like to participate in future productions that integrate some of these seeds into new global arts/performances. A brief distillation:

I would love to participate in groups via cyber space to feed or review or suggest content from the 15 Global Challenges to arts/media groups…Flash Mob Art against poverty using Twitter and Facebook…I would like to participate in these global challenges… What we need are exchange programs, mentoring and coaching for artists, creative producers of contents, students and children. We need labs, beta areas, digital and virtual test-playgrounds to apply new ideas and concepts, to exploit the digital power to bring people closer together. Thanks for this great Real Time Delphi which really makes a difference!

Conclusions

As creativity and self-expression are inherent desires for all individuals, we will continue to see trends toward more immersive experiences, audience participation, fan-based contributions, smart gaming, socially beneficial storytelling, innovative ways to use the newly emerging technologies, and DIY and independently created intellectual properties that are developed through a more technically democratized system of distribution. New screens and portals for delivering stories, media, and interactive experiences will continue to emerge, and a more "transmedia" approach to narrative development and distribution is emerging to exploit these new platforms.

Social media and digital media will continue to catalyze the expansion of "free speech" and be a tool to expose human rights violations and unjust governments, as we have seen in the Arab Spring-Awakening this year (using Twitter). With the proliferation of more media than ever, there should be an emphasis on media literacy being taught from an early age up into adulthood.

Traditional media will continue to evolve into and through new media platforms, but a solid "story" and narrative will always be at the heart of these experiences. Transmedia strategy will become the norm during development stages of media content, as the core intellectual property or story worlds will be developed with new distribution models and emerging portals of entry offering media makers and artists new access for expression and concertizing their concepts and creative properties. Cinemas will likely evolve to become local place-based transmedia distribution centers allowing greater immersion in visitors' favorite stories through immersive cinemas, video game tournaments, immersive gaming pods, and regional alternate reality games. Electronic publishing and the nature of the book will evolve, and new authors will have the ability to become popular based on the quality of their content and their marketing efforts. Niche subject matters and audiences will proliferate as the Internet offers exposure for any topic or experience.

Games will continue to become a more popular platform for education and other applications in addition to serving as an entertainment art form.

As creator of the serious game EVOKE, author Jane McGonigal writes in her new book, *REALITY IS BROKEN– Why Games Make us Better and How they can Change the World:*

Gamers want to know: Where, in the real world, is that gamer sense of being fully alive, focused and engaged in every moment? Where is the gamer feeling of power, heroic purpose, and community? Where are the bursts of exhilarating and creative game accomplishment? Where is the heart-expanding thrill of success and team victory? While gamers may experience these pleasures occasionally in their real lives, they experience them almost constantly when they're playing their favorite games.

The truth is this: in today's society, computer and video games are fulfilling genuine human needs that the real world is currently unable to satisfy. Games are providing rewards that reality is not. They are teaching and inspiring and engaging us in ways that reality is not. They are bringing us together in ways that reality is not.

The traditional, folk, indigenous, and classical arts will also benefit by combining new media technologies with traditional art forms in an effort to reach diverse new and global demographics. As an example, note that opera has reached new audiences through the digital distribution into theaters and the video projection accompanying it; classical music is also becoming popular and drawing in more diverse and younger audiences. With mobile technological advances, economically challenged individuals and cultures are also able to create content using mobile video cameras so that their culture and stories can also be expressed in a global audience space.

Music, media, and the arts remain powerful tools for bridging cultures, sharing diverse values and perspectives, addressing conflict, educating, and ultimately creating new visions for a more evolved and peaceful civilization.

As we continue to explore and create these new emerging technologies and hybrids, let us combine this innovation with intelligent awareness, a passion for meaning, and global collaboration toward a more enlightened humanity. Let us discover and create a new "opera" in the spirit of a twenty-first-century Camerata and a better future for all.

courtesy of: http://upload.wikimedia.org/wikipedia/commons/6/62/Latin_America_regions.svg

5
Latin American Scenarios 2030

Between 2010 and 2030 most countries of Latin America are commemorating 200 years of independence in multiple bicentennial celebrations across the region. As these countries look back over their first two centuries, it seems appropriate to take this opportunity to explore some possible futures for Latin America. The last 200 years provide a basis for thinking about the next 20 years.

Scenario analysis has become popular in many places since the middle of last century. In the twenty-first century, many entities—from companies, cities, and countries to regions—are using scenario analysis to help them make policy decisions. However, no major Latin American scenarios have been developed within the region during the last few years.

Latin America is a major world region encompassing Mexico, the Caribbean, Central America, and South America. Most countries in the region became independent following the French invasions of Portugal and Spain by Napoleon in the early 1800s. The region was then usually called Ibero-America, a term still used mainly in Portugal and Spain, but Napoleon III supported the term "Latin America" during the French invasion of Mexico in the 1860s. The term "Latin America" was also sometimes applied to include other French former colonies from Canada to the Caribbean and was used by some intellectuals who linked the region to the linguistic roots of French, Portuguese, and Spanish in Latin. Thus, linguistically, Latin America is an even larger geographical area that could also include some parts of the U.S., which is now the second largest Spanish-speaking country in the world (after Mexico but ahead of Spain). "Latinos" or "Hispanics" today represent close to 13% of the U.S. population, and they are the single largest U.S. minority.

The population of conventional Latin America—from Mexico and the Caribbean to Argentina and Chile—has grown considerably during the last two centuries after having been significantly reduced in the decades immediately after the arrivals of Europeans, who brought new diseases to the region, unknowingly decimating large groups of indigenous groups. The Latin American population stood at around 576 million in 2010 and is expected to stabilize in the second half of this century at over 730 million. In a global context, the populations of the European Union (currently with 27 members), Japan, and Russia are already declining. The population of China will also begin shrinking in the 2030s, and India will then overtake China as the most populous country in the world (see Figure 35). The population of Africa will keep on rising until the end of this century, when it is expected to stabilize as well.

In terms of economic development, Latin America was a relatively wealthy region at the start of the nineteenth century. In fact, some parts of Latin America were richer than the nascent United States. The Dominican Republic, Mexico, and Peru had universities almost two centuries before Harvard was founded. Haiti was a very wealthy colony in 1800, richer than many parts of the U.S. then. Latin America was at par with most of Europe, and it was richer than Africa, China, India, and Japan. In fact, even at the beginning of the twentieth century Argentina was one of the 10 wealthiest countries in the world, and many poor Chinese and Japanese immigrated to richer Latin American countries like Brazil, Mexico, and Peru. By the beginning of the twenty-first century, however, Latin America fell behind, and many countries in East Asia had overtaken it. If current trends continue, China will overtake Latin America in terms of GDP per capita in the 2020s (see Figure 36).

Figures 35 and 36 show single-point projections for the population to 2050 according to the UN and the GDP per person to 2030 extrapolating the 2011–15 forecasts by the IMF. Population forecasts are easier than GDP forecasts since they are smoother and more predictable, as the curves in Figures 1 and 2 exemplify. This is also why the UN has population forecasts to the year 2050 (in fact, there are even demographic projections up to the year 2300), but the IMF only has five-year forecasts, which have been extrapolated here to the year 2030.

Figure 35. Comparative evolution of population (linear scale): Historic and projections, 1800–2050

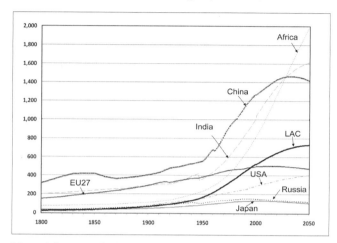

Note: The population projections correspond to the medium variant of the UN. LAC refers to Latin America and the Caribbean.

Figure 36. Comparative evolution of GDP per person (logarithmic scale): Historic and projections, 1800–2030

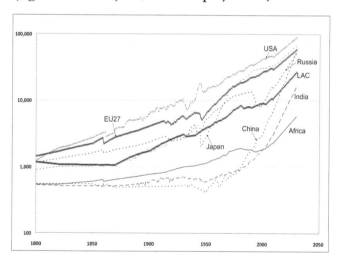

Note: The GDP per capita projections are an extrapolation to 2030 using the same growth forecast for 2011–15 as the IMF.

Table 7 compares Latin America and other major countries and regions in the world according to their land area, population density, population forecasts to 2050, and GDP forecasts to 2030 (both total and GDP per person). China is developing fast, and if its rates of growth continue it will overtake Russia and Latin America, while India will also get closer to Latin America by 2030.

Thinking beyond the GDP, and certainly including more than economics, we can use a STEEP (society-technology-economics-ecology-politics) analysis in order to consider other variables. The GDP is an important variable but certainly not enough, and an analysis using only the GDP is too simplistic. Therefore, we can also consider the Human Development Index developed by the United Nations Development Programme and other variables. Table 8 shows some of the variables included during the Delphi Survey for this Latin America 2030 study. It is useful to analyze the "best" and "worst" values for each variable, both in Latin America and in the world, as well as the corresponding average values.

Table 7. Latin America in the global context

Country or Region	Area (million km²)	Population density (people/km²)		Population (million)		GDP (PPP, billion US$ 2010)		GDP per person (PPP, thousand US$ 2010)	
		2010	2050	2010	2050	2010	2030	2010	2030
Africa	30.222	37	66	988	1,998	2,348	11,686	2.376	5.849
China	9.641	139	147	1,337	1,417	10,051	80,097	7.518	56.526
India	3.287	359	491	1,181	1,614	3,887	26,418	3.291	16.368
Japan	0.378	336	270	127	102	4,296	6,878	33.828	67.434
Russia	17.075	8	7	141	116	2,229	6,087	15.807	52.478
USA	9.827	32	41	312	404	14,705	36,373	47.132	90.034
EU27	4.325	116	109	501	473	15,213	28,016	30.367	59.230
LAC	21.070	27	35	576	729	6,444	19,650	11.188	26.955
World (land)	148.940	46	67	6,909	9,150	74,004	240,246	10.711	26.256

Notes: The numbers do not add up to world total since not all the countries/regions have been included. The population projections correspond to the medium variant by the UN, and the GDP per capita projections are an extrapolation to 2030 using the same growth forecast for 2011–15 as the IMF.

Table 8. Comparative best and worst cases for international indexes, 2010 (economics-society-ecology-politics-technology)

Variable/Indicator/ Index	World Worst	Latin American Worst	World Average	Latin American Average	Latin American Best	World Best
Society: HDI (from 0 worst to 1.000 best)	0.140 (Zimbabwe)	0.404 (Haiti)	0.624	0.704	0.783 (Chile)	0.938 (Norway)
Technology: E-Readiness Index (from 0 worst to 10 best)	2.97 (Azerbaijan)	3.97 (Ecuador)	4.30	5.40	6.49 (Chile)	8.87 (Denmark)
Economics: GDP per person (PPP, thousand US$ 2010)	340 (D.R. Congo)	1.121 (Haiti)	10.711	11.188	19.600 (Puerto Rico)	88.232 (Qatar)
Environment: CO_2 emissions (tons/person)	55.5 (Qatar)	6.0 (Venezuela)	4.6	3.7	0.2 (Haiti)	0.0 (Mali)
Politics: Corruption Index (from 0 worst to 10 best)	1.1 (Somalia)	2.0 (Venezuela)	3.3	3.6	7.2 (Chile)	9.3 (Denmark)

Notes: The best and worst values correspond to the latest information of the countries with available data in 2010. The Latin American and world averages are based on population-weighted values.

Considering multiple variables gives a broader spectrum to study the future of Latin America, both in terms of itself and also in comparison with other regions and countries around the world. Latin American nations have fallen behind several other countries in the last 200 years. What could happen in the next 20 years? Will the situation in Latin America become better or worse? In fact, different scenarios actually consider both possibilities. Diverse variables have to be analyzed in order to avoid the worst and to reach the best alternatives.

Methodology

In 2009 The Millennium Project initiated a multi-year study about the future of Latin America. This coincides with the expected multiple bicentennial independence celebrations throughout the region. The first phase of this study consisted of a Real-Time Delphi survey during 2009–10; the second RTD, run in 2010–11, was designed to integrate the results of the previous one in the form of 2030 Latin American scenarios.

In the earlier study, The Millennium Project Nodes in Latin America designed an RTD to collect judgments from knowledgeable individuals about the likelihood and significance of diverse international and regional developments that might affect Latin America over the next 20 years and about the potential course of variables important to the region. The other Nodes of The Millennium Project around the world also helped identify additional experts to give an "outsider" view of Latin America. The RTD was distributed in English, Portuguese, and Spanish and had a total of 92 questions, divided into:

- International Developments (questions 1 to 38)
- Geopolitical Influences (questions 39 to 52)
- Latin American Developments (questions 53 to 82)
- Scenario Axes (questions 83 to 87)
- Main Variables (questions 88 to 92)

More than 550 people from about 60 countries participated during a seven-week period. About 30% identified their gender as female. By country, the top participation was from Brazil at 19%, followed by Argentina at 15%, Mexico at 13%, Peru at 13%, and the United States at 12%. The Real-Time Delphi helped identify some developments with high likelihood and high significance (called "good bets") and some others with relatively low likelihood but high significance (called "surprises"). Additionally, the experts corroborated the "rise" of China and the positioning of Brazil as the most influential Latin American country. The results of the Delphi survey (www.millennium-project.org/millennium/RTD_LA2030/LatinAmerica2030DelphiSurveyLong.pdf) and a compilation of the answers by the Delphi participants (www.millennium-project.org/millennium/RTD_LA2030/LatinAmerica2030DelphiSurveyAppendix.pdf) can be seen online.

After the conclusion and review of the RTD results, the Latin American Nodes of The Millennium Project decided to create four scenarios for Latin America 2030 using a techno-economic axis and a sociopolitical axis. Using these standard but simple axes allows the design of a scenario matrix that can be easily visualized. First of all, scenarios allow consideration of many more different possibilities than a single-point forecast. This is a major advantage of the use of multiple scenarios—that is, they increase the range of possible futures to be analyzed, as shown in Figure 37.

Figure 37. Single-point forecast versus scenarios

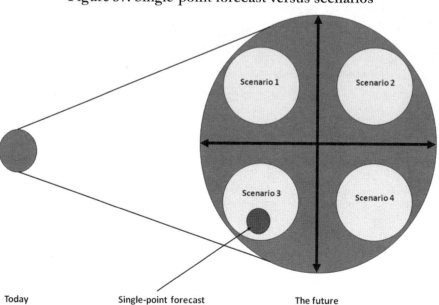

Today   Single-point forecast   The future

Furthermore, the stories behind the scenarios help identify additional factors and broaden the vision of what might be possible, even if not very probable. Finally, some scenarios might even reflect single "wild card" events (or the "surprises" identified in the previous RTD) that might have high impact or significant consequences.

Figure 38 shows the four scenarios created using the techno-economic axis in the vertical position and the sociopolitical axis in the horizontal position:

- Scenario 1 – "Mañana" is Today: Latin American Success
- Scenario 2 – Technology as Ideology: Believers and Skeptics
- Scenario 3 – Region in Flames: This report is SECRET
- Scenario 4 – The Network: Death and Rebirth

Each scenario builds upon the information gathered during the Real-Time-Delphi and the direct feedback received from other Millennium Project Nodes.

Figure 38. Scenario matrix

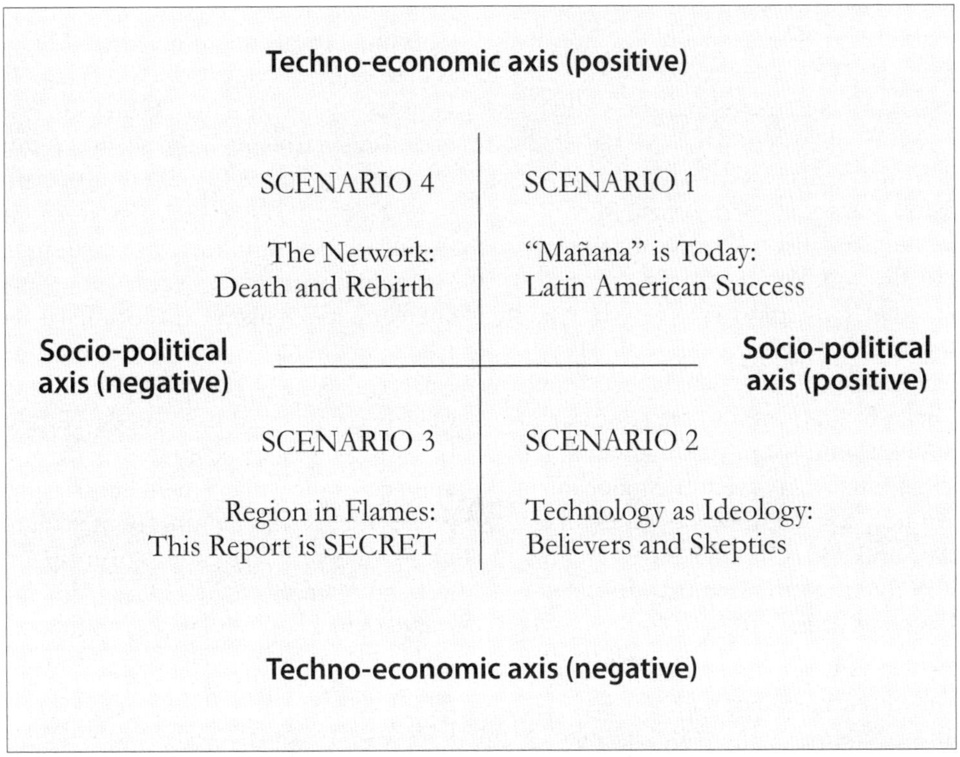

	Techno-economic axis (positive)	
	SCENARIO 4	SCENARIO 1
	The Network: Death and Rebirth	"Mañana" is Today: Latin American Success
Socio-political axis (negative)		**Socio-political axis (positive)**
	SCENARIO 3	SCENARIO 2
	Region in Flames: This Report is SECRET	Technology as Ideology: Believers and Skeptics
	Techno-economic axis (negative)	

Different Latin American Nodes of The Millennium Project coordinated each scenario, including several "fill-in-the-blank" questions that participants were asked to complete in the scenario drafts. Based on their feedback, each scenario was rewritten to best incorporate all additional input.

Scenario 1

"Mañana" is Today: Latin American Success

From Important Speeches of the Month, June 2030

This is a keynote speech presented by Javier Bolivar, Chairman of the Latin American and Caribbean Union to the Assembly of African Nations in response to their question: "Upgrading Development: How did Latin America do it?"

It is a real pleasure for me to be with you today to speak about our experience and policies, and perhaps a bit of good luck, in achieving our present state of development. I hope that nothing of what I will say might sound as bragging, and I offer this hoping that it will be useful to you and your nations.

This year, 2030, began very well for the world in general and for the Latin American and Caribbean countries in particular. The State of the Future Indexes for the countries in the region show continued progress. This proves what can be achieved when there is international political will to trade some sovereignty for the common good. Brazil, Mexico, and other countries have made important progress in education and also in fighting poverty, crime, and drugs. Living standards are improving very fast, and social disparities have been greatly reduced. Many new technologies, some of them developed in Latin America, have been fundamental to the educational renaissance and economic boom that the region has been experiencing. Finally, most Latin American countries can proudly say that they are joining the ranks of the developed world, which has been greatly transformed with the incorporation of other major nations like China and India.

Let's go back to the UN meeting in 2015 that reviewed the Millennium Development Goals established at the UN Summit in 2000. During the UN review, the Group of Latin America and Caribbean Countries announced a new plan for the region: to reach "developed status" in another 15 years, by 2030, since the MDGs for the region have been largely met. This was not just a diplomatic announcement; it was fully backed by the civil societies of every country, by local and regional NGOs throughout Latin America, and by the private sector, universities, and governments. Coincidentally, the eyes of the world were on Latin America with the very successful 2014 FIFA World Cup in Brazil and the scheduled 2016 Olympic Games in Rio de Janeiro. Former Brazilian president "Lula" da Silva had famously said that "God is Brazilian"; well, maybe it was actually becoming true for Latin America. Thus, Brazil functioned as the international showcase of a new Latin America striving for more and better development. Traditional measures of economic success, such as GDP, had been superseded in the

wake of massive global restructuring, leaving many Latin American countries to reinvent prosperity around new metrics of value, connectivity, resilience, influence, and happiness, stimulated by the State of the Future Index processes.

Most Latin American governments really made a concerted effort to advance toward political and economic union under enlightened leadership. Chile set the pace, becoming the first large Latin American country to have reached "developed status" by the early 2020s. Chile was the second Latin American country to join the OECD in 2010, after Mexico in 1994, but Chile achieved increasing prosperity during longer periods thanks to more open and stable public policies. Larger countries like Brazil and smaller nations like Costa Rica have also followed paths similar to Chile's. Even after the terrible earthquakes of 2010 and 2020, Chile managed to quickly recover and keep on developing very fast. It became a regional example of progressive political and economic achievements under alternating governments. In fact, the country managed to keep improving very fast under governments from the left and from the right. Chileans joked that they did not care much about left or right, because they just wanted to move forward and upwards. And talking about upwards, Chile, together with Argentina and Brazil, participated in a joint unmanned mission to Mars, which also had collaborators from Asia and Europe. Peru decided to specialize in biotechnology, giving more added value to its Andean and Amazon biodiversity. By 2030, a good part of all the vegetables and fruits sold worldwide came from varieties developed in Peruvian biotechnology labs under public-private alliances, respectful of local intellectual property rights. Mexico also had great advances in biotechnology, including important developments with several corn varieties and the unique nopal cactus.

The 2020s were a time of very rapid development in Latin America, much faster than expected at the turn of the century, and by 2030 the majority of Latin America's 730 million people have a good living standard. Most Latin American countries managed to reduce poverty by improving the quality of education for all, which allowed these countries to reduce the gap between public and private education. Functional illiteracy was eliminated and nearly all people had access to tele-education systems. Many people had access to their own Internet-based businesses, looking for markets instead of looking for jobs. The new education systems stopped the brain drain and promoted human success by investing in human development, including fairness in distribution opportunities, goods, and services, by reducing gross inequalities of income, wealth, and power. Most Latin American nations also managed to improve the health and well-being of their citizens while creating sustainable economies that set an example for the rest of the world.

Latin America's successful development has transformed it into the most prosperous and happiest region, even ahead of the European Union and the United States, thanks to the very fast internal and external development of the region. Latin Americans enjoy the healthiest environment in the world: large open spaces, unpolluted air and water, clean beaches, sustainable rainforests, and even archaeological sites combined with modern adventures. In fact, the world's most visited place has become the new Disney Entertainment Park at Cuiabá, Mato Grosso, in Brazil, close to the geographical center of South America. A fundamental achievement of Latin Americans has been to "latinize" the rest of the world by promoting the Latin American lifestyle. The "happy style" of Latin America was exported to other parts of the world thanks to the "happy" actors, artists, dancers, musicians, singers and writers originating from Argentina to Venezuela, from Brazil to Mexico, from Colombia to the Dominican Republic, from Chile to Peru, from Bolivia to Cuba. Even the Latin American gastronomy, including many exotic drinks and typical foods, has become a recognized symbol of a happy lifestyle.

Caudillos, military coups, dictators, guerrillas, and terrorists are now part of history, and strong democracies are the norm in the new Latin American and Caribbean Union, which has adopted a regional currency to favor its flourishing regional and international trade. The largest market for Latin American exports is now China, followed closely by the United States and then Africa, Europe, and India. The growth of Latin American internal trade has been impressive, and it now complements a much diversified external trade balance with all of the other major trading blocks.

The creation of the Latin American currency was fundamental to consolidate the financial stability and strength of the regional monetary system, since all major international economic blocks also used common currencies linked through global electronic exchanges. Our "Structural Development Fund" has allowed the least developed countries of the continent to stabilize and "balance out" with respect to the most prosperous ones.

Now, in 2030, the Latino and Hispanic population in the United States is by far that country's largest minority, with over 70 million people today and still growing. This has been reinforced by the large historical Latin American "diaspora" in the United States, and to a lesser extent in Canada. The United States has thus become the second-largest Spanish-speaking country in the world, after Mexico, but ahead of Argentina, Colombia, or Spain. In fact, during the U.S. elections of 2028, Juan Pérez was elected as the first Hispanic president, following his two successful periods as governor of California. The election of the first Hispanic U.S. president allowed the United States to see Latin America as a close neighbor rather than the U.S. backyard, thus reducing tensions and improving bilateral relations. President Perez also advocated that the United States join the more vigorous Latin American and Caribbean Union that was created to consolidate all the previous social, economic, and political regional groups throughout Latin America and the Caribbean.

Not only has the international trade of Latin America increased substantially, but so has the quality of its products. "Made in Latin America" became a sign of excellent quality and is also a label indicating top environmental and social responsibility. There is a new "continental nationalism," and a product from a Latin American country is usually preferred over products from Asia, Europe, or the United States. The concept of "Latin America Motherland" is also a reality, and there is free mobility across the Latin American borders. People can circulate freely without passports or any documents, except for some special areas that use retina scanning for identification.

In this discussion I should not ignore the great changes occurring elsewhere in the world, which helped create a climate of peace. The reunification of North Korea and South Korea in 2020 and the International Peace Treaty of Jerusalem in 2022 were very important events that paved the road toward a more peaceful world. Latin America has been a nuclear-weapons-free zone since the Tlatelolco Treaty in 1967, and most Latin American countries eliminated their armies in the 2020s, following the example of Costa Rica, which had done so in 1948. This anti-war movement started in 2022, when Brazil, celebrating the bicentennial of its independence, declared that the only Latin America war was against drugs, poverty, hunger, and poor education. The whole region established a continental plan to transform its armies into national guards devoted to defending people against natural disasters.

The eradication of illegal drugs was achieved as a result of the reduction of European and U.S. drug consumption. The victory over the drug cartels created a wave of hope and optimism over the whole continent, from Canada to Argentina and Chile. In a more peaceful world, with a non-belligerent United States and separated by oceans from other world regions, Latin American resources that previously went for defense are now devoted to education, arts, science, and technology. Obviously, there are always risks, but the growing prosperity and more transparent governance in Latin America and the rest of the world make it very unlikely that internal or external wars will occur, just as a European war is unthinkable today among the members of the European Union following its creation in the twentieth century.

Breakthroughs in science and technology around the world also played a role. No matter where these advances originated, they spread quickly throughout the planet. The World Trade Organization, World Intellectual Property Research Organization, and Internet 7.0 helped ensure that knowledge moved fast from country to country. Technology keeps getting better, cheaper, and faster. It is now estimated that almost all Latin Americans are continuously connected to Internet 7.0 with their mobile jewelry and clothing nano-telecomputers. Synergies among nanotechnology, biotechnology, information technology, and cognitive science (commonly known as NBIC technologies) have dramatically improved the human condition by increasing the availability of energy, food, and water and by connecting people and information anywhere, anytime. The positive effect has been to increase collective intelligence and to create value and efficiency while lowering costs. However, some people complain about too much technology and unintended consequences, such as over-reliance on technological solutions. Most of our institutions try to guard against the notion that "technology will fix anything."

A key facilitator of this important process of rapid development was the creation of the Latin America University in Panama. Located in the former "Ciudad del Saber" (City of Knowledge) on the grounds of the old U.S. Army facilities, and fully accessible in cyberspace, LAU obtained financial resources from all the Latin American countries according to their respective wealth. The new regional tele-collaboratories at the LAU campuses received many Latin American researchers and scientists who returned from other important labs around the world, in a reverse diaspora that brought the "brains" and experience back to the region. LAU has now become one of the most important R&D centers in the world, and along with other regional universities it is accelerating the process of technology transfer and innovation across Latin America. For the first time, many Latin American universities began appearing in the top ranks of the best world institutions.

The acceleration of technological development, successful economic policies, and national education goals to increase human intelligence have opened the door to continuous and rapid economic growth. The NBIC technologies are proving to be the key to a very bright future, in which machines work in increasingly efficient ways so that the cost of goods continues to plummet and tremendous wealth is created faster and faster for everybody. All basic necessities, as well as intellectual and physical luxuries, can be accessible to even the poorest people, in what some experts call the "post-scarcity" society where everything is abundant and cheap. Space exploration, artificial intelligence, and robotics are close to a take-off point that some experts refer to as a "technological singularity," probably during the late 2030s or early 2040s.

Some experts talk about the possibility of a time of smarter-than-human machines, although others think that the "technological singularity will always be 20 years off in the future because the definition will keep changing." Still, a few others believe that another version of the singularity could arise from the integration of human and other kinds of life-forms, which would be fauna and flora consciousness. The basic line of progress will be a breakthrough in genes that control our communication interface. In this version of a benign singularity, the Amazon will transform to a hub of live intelligence infrastructure. So preserving it now would mean that in the distant future these areas will become strategic assets for an emerging global power like Latin America.

Meanwhile, amazingly enough, Moore's Law still seems to hold true, and computers continuously become faster, smaller, and more powerful—now integrated into much of the built environment and clothing. Quantum computing, 3D circuits, and architectural innovations have given new life to Moore's Law. The largest computers now have more transistors than humans have neurons in their brains. Artificial intelligence now augments human intelligence, as a common experience. We are now on the threshold of incredible scientific developments, such as humans being transformed into more advanced life forms: transhumans and posthumans. In fact, the first cyborgs and clones are already becoming accepted and normal in some societies, and their numbers are increasing faster than those of the so-called naturals. Biological evolution, which is slow and erratic, is being overtaken by technological evolution, which is faster and directed.

Many humans will never be the same, and some people worry that the very nature of "humanity" and what it means to be uniquely human is blurring and losing distinction. Creating new life forms or modifying them for human purposes continues to cause shockwaves across the theological landscape, eroding one of the most durable institutions of Latin America: the Catholic Church. Now the very faith of Latin America is being challenged. This worries some people, but I think it is an incredible opportunity. Yet, just as some Andean highland communities and Amazon tribes still refuse any technology today, some other Latin Americans are still afraid of more scientific advances. A common decision has not been reached, and perhaps it will never happen, but individual choice is guaranteed by most governments, and choice includes the relationship to human-changing technology. Yet we are alert to the need to preserve regional identities, traditions, and cultures, as established in the 2028 LACU constitution.

While some people debate what it means to be human in general and Latin American in particular, the 2030 Mexico City World Expo has been incredibly successful, and when it closes on October 31st it is expected to have been the largest international gathering in human history. So far close to 50 million human tourists and over 100 million robots have already visited the fair grounds, conveying their experiences through virtual reality to their families and friends. Cyber-visits have averaged over a million per day, and both humans and robots have seen the great advances brought by science and technology in order to create a better, cleaner, and more peaceful world. One of the most visited pavilions was the one displaying science and technology in Latin America, including major breakthroughs in the control of aging and enhanced rejuvenation processes, as well as significant advances in quarktronics, leptronics, femtotechnology, and the planned human mission to Mars by the Latin American Space Federation.

In the meantime, inequality—in its broadest sense, as social, technological, economic, environmental, and political—has been the single toughest issue to overcome. But if the trends of the last 20 years are any indication, then the future will be very bright for the region during the next two decades. Against many odds, international cooperation and good public policies have allowed a virtuous cycle of development and improving conditions for most people. Therefore Latin America will happily celebrate not just two more decades, but hopefully two more centuries of progress and prosperity. A new civilization is being born, and Latin America is not missing the train this time. The future has finally arrived: "mañana" is today!

Scenario 2

Technology as Ideology: Believers and Skeptics

This is a true record of the June 20, 2030, debate at the Latin American University forum on the role of new technologies in building a new Latin America. The speakers had five minutes each to make their primary points for and against a positive role for technology, followed by rebuttals. Speaking for the negative, Dr. Juan Bosque, and for the positive, Dr. Francisco Arbusto. Based on a coin toss, the negative will begin.

Dr. Bosque (negative):

Although Latin America has some NBIC-based technologies today, these new capabilities haven't accelerated our social and economic development. Applying external technologies without understanding both their potential and their downside leads to an inconsistency in their application. It is certainly evident that living standards in our region have improved even more than we might have expected 20 years ago, but it is equally apparent that they have resulted in further concentration of wealth, have raised expectations which can't be fully met, and have broadened the social and economic gap between classes.

Focus on short-term profits, and other ills that led to the world financial crisis in 2008, lingered in Latin America, preventing what could have been a more equitable technological development. What we needed were public policies for solving structural problems of the politico-economic system for improving education and for accelerating socioeconomic development.

In the 1980s, Argentinean futurist Horacio Godoy anticipated one of the key problems of technological development in what he described as the USTED syndrome (the underdeveloped use of developed technologies): the mismatch between expectations of technology and the actual contributions. What happened? Many groups across the region claim that the production based on new technologies was driven by the motivation of short-term profit and ignored the human, social, and environmental costs (such as planned obsolescence) that they would bring. Consider, for example, the expanding use of harmful agrochemicals, pharmaceuticals designed to cure diseases but not to promote health, and advanced weaponry. Some technologies threaten human personality; democracy is still perceived as a euphemism that covers the role of "corporatocracy" as the force from which the politico-economic power really derives.

New technology merely excites unwarranted consumerism and may create new needs without fulfilling some basic needs. What is the rationale of having people use smart phones 5.0 without proper clean water and sewage systems? Products have a shorter and shorter useful life, as due to new technologies they become increasingly sophisticated and the need for renewal is endless. Advertising increases the wish for new products, but only a small percentage of the population could afford them.

Products based on new technologies only give the already powerful even more power, while they still use outdated strategies of war and follow the agendas and styles of western democracies. Other problems are focused especially on land use. Think about GMO-based foods: the seeds are supplied by monopoly agribusinesses, so they set the prices; yes, the harvests may be larger, but where are the benefits for the rural small farmers? What is the biotechnology that besides larger crops also provides employment for rural people? Economies based on cooperation and shared ownership of productive goods and resources have made some progress, but the ideal of converting the gains of productivity into reduced work and free time for the working society has yet to be realized. We had hoped that increased productivity could be converted to improved human qualities, attributes, and potentials; instead, we find that improved productivity equals higher unemployment.

Dr. Arbusto (positive):

You make some good points, Dr. Bosque, but you're wrong about much of your assumptions. The benefits of technology far outweigh the risks, and the future status of the nations and people of our continent depends on them. While we still have social crises—including poverty and marginalization—things are much better today than they were 20 years ago.

Take the global financial crisis of 2008 and the subsequent world recession: Latin America was not affected as deeply as many expected. The number of people making a living online has exploded over the past 20 years; people are finding work and how to learn and earn in new ways. This modernization in job creation occurred in a context of growing tension between creativity and imitation. Many local "traditional" jobs were sacrificed in the interest of competing on the global market, but the capabilities that new technologies offered provided a means of employment that did not exist before. It is true, nevertheless, as you suggest, that some new technologies were used merely as a "sign of modernity" without creating new ventures or generating sustained overall economic growth or development. Nevertheless, advanced technologies were vital key improvements: environmental problems such as pollution in major urban centers and industrial areas have been partially controlled; pollution and falling water tables have been curbed by improvements in the treatment and disposal of garbage and more attention to rivers and drainage systems in water projects. Tele-education and tele-medicine have made major contributions to improving the quality of life for many people. We have flexible emergency lodging and other response capabilities to natural disasters, we have earthquake-resistant buildings, and we have cheap and more efficient energy.

Dr. Bosque (negative):

If new technologies had not been a source for improving our lives, what have been the causes? Early in this century, outside corporations took control of the natural resources in the region. Global corporations have managed to transfer much of the costs of environmental consequences to nation states, as was the case of infrastructure development in industrial society, seriously compromising future generations. Now, global corporations with great financial power are attempting to do the same with biological resources—to extract and patent elements of the rich regional biodiversity (plants, animals, microorganisms, and all kinds of living matter and components of biological material). Biodiversity is at risk of being appropriated by corporate globalism. This violates the very sovereignty of countries.

Moreover, no consensus has been reached for helping countries in the region to act as a bloc against the abuses of multinational corporations, which therefore could operate indiscriminately because they were creating jobs for the region. Countries by themselves are unable to outline a regulatory framework governing the use of these biological materials in all Latin America.

Dr. Arbusto (positive):

Let me remind you that there were a series of crises of governance by 2020; charismatic but egocentric leaders of the past who were supported by patronage and demagogy have fallen due to public pressure. Before the crises erupted, there were serious attempts to control the vast resources for government procurement and the mass media (such as Internet 7.0), but like-minded people across the subcontinent (using one of the technologies you deride) met in cyberspace and formed transnational power blocs. Communications were in cyberspace but the confrontations were in physical space. Political science theorists saw the events as self-organization in the midst of a chaotic system. The outcome was clear from the beginning. This technology brings the world to us and us to the world. It permitted people to participate in the rewriting of constitutions, it helped identify and train new leaders for replacing the old guard, and it consolidated representative democracy. This capacity is still young and the capabilities are still partial and limited. Real democracy is still impeded by reminiscences of the old economic system in which rules were set from the top down; hence, this environment is still new and difficult for voters.

When the dust settled, by 2025, we had improved our sociopolitical systems of governance while the old values still largely prevailed: family solidarity, religious values, and international charities. We found improvements in almost all areas: income, unemployment, literacy, rich-poor gap, but perhaps the most important use of the new online technologies is in education. People of all ages had all the humanly available information at their fingertips.

Dr. Bosque (negative):

Ah, there's the problem: people who need access are deprived of it by cost, time, and lack of knowledge about how to use it. We may have information but what we need is wisdom. Without an educated public, it is difficult to make democracy work. What we need is a dynamic balance of peace and justice. In addition, we still must work hard to provide the quality education you and your technologies promise us.

Let me point out that new important issues have appeared and old ones have worsened, such as land use and management policies, housing shortages, regressive taxation, concentration of corporate ownership, farm subsidies, and growing bribery, corruption, and organized crime. The scarcity of medium- and long-term programs with support from local leaders worsened these problems and the social unrest.

Dr. Arbusto (positive):

You know, despite our regional problems, the global situation has improved dramatically. Perhaps this has to do with the new soft technology of improved decisionmaking. We have seen agreement between Taiwan and China, and Korea's reunification in 2020, and the peace treaty of Jerusalem in 2017. The United States effectively repealed the Monroe Doctrine in 2028, abandoning its self-assigned "protectorate" role and the use of military intervention in the region in support of what appeared to be its national interests. With fewer conflicts between states, Latin America has substantially changed its concept of defense and security. Our single regional army, under the Latin American Caribbean Union, was formed with the primary role of assisting the victims of natural disasters, which in the last two decades exceeded by far those of historical wars. This bold security policy changed the military rationale from border conflicts to responses to natural hazards: earthquakes, tsunamis, landslides, and floods. The problem, of course, was how this armed force would be commanded. We looked at many possible solutions:

- A supranational army under the direction of a mature regional military
- A Latin-American Council of Defense integrated by all countries in the region and with a rotating presidency
- An institution independent and apolitical, with equitable representation from every country
- An institution like the North Atlantic Treaty Organization, to which Latin America nations can contribute, rotate to choose institutional leader, and select permanent and non-permanent committees as in the United Nations
- Rotation of command among the contributing countries, or a non-military top command
- A coalition that includes representatives of all countries, with the participation of experts and professionals in various fields, such as doctors, engineers, sanitarians, planners, nurses, scientists, politicians, psychologists, volunteers, and so on, dissolving the old feature "military" and turning this into a common core structure for humanitarian purposes.

I won't say that these approaches were technologically

inspired, but without advanced technologies they would have been more costly and less practical.

Dr Bosque (negative):

Dr Arbusto, you can't claim that new technology was the main cause of reduced corruption. There is a long list of long-term policies and strategies that made a difference, such as increased transparency; international cooperation on arresting drug dealers and money launderers; public consultation to restore people's confidence in political parties and institutions; improved laws and implementation of public disclosure of government officials' financial affairs; support of civil society and culture; establishment of anti-corruption networks and blacklists of corrupt officials (convicts of corruption were banned for life from holding public office); judicial review with reform of criminal laws and procedural practices (such as dilatory recursive accelerating trials); and citizen networks to oversee procurement, purchasing, evaluation, and execution of contracts with public accessibility.

Dr Arbusto (positive):

Hold on a minute, Juan! Don't you think that new techniques of forensic accounting helped find money launderers? Don't you think that tele-classrooms helped bring excellent education to people who were denied it before? Don't you think that new media of public communication helped create the civil culture you refer to? Latin American artists and entertainers used new forms of cyber media to increase the public's courage to demand better governance. An important aspect was the reintroduction of the so-called *juicio de residencia* (impeachments) of the colonial times, ending the reign of highest government officials. Any citizen could complain about crimes committed during an official's tenure, potentially concluding with the imprisonment of the corrupt officials and their corrupters. Anti-corruption programs also included tax reforms based on equity, efficiency, fiscal responsibility, and mechanisms to stop tax evasion, as well as taxation and accountability mechanisms for foreign corporations operating in Latin America.

I'm sure you know that we were able to trace each individual peso and real with an electronic signature; this was of enormous benefit in the fight against money laundering and corruption activities. Health services, education, housing, and security improved as well.

Dr. Bosque (negative):

But how about energy and national economies? National energy policies designed to protect the

environment were ineffective; they seemed to be designed and brokered by the corporations involved in energy production and distribution. The national economies of Latin America are still fragmented: the more competitive and larger business sectors—which were part of the global economy—received most of the attention of the new governments, which applied economic and financial policies to better protect them. A second sector, oriented toward the domestic market, continued its difficult development, continuously threatened by the opening of the economy, dumping tactics, and smuggling. The third sector, the informal economy, continuously expanded and became an even stronger contributor to national well-being. It consisted of people who were unemployed in the first or second sectors. They developed strategies for survival outside of basic health services, education, and even personal dignity; many created online services and found markets around the world that are now totally connected through the Internet.

Dr. Arbusto (positive):

OK, people were able to develop those online businesses thanks to our computer literacy programs and excellent simulators that helped them develop management, business, and interpersonal skills, as well as learning new languages and get basic training in virtually any domain. These programs included the active collaboration of universities and colleges for the implementation of e-commerce, market organization through networks, distribution logistics, inventory management, creation and use of innovations such as electronic funds transfer, and the establishment of labor organizations for government backing the entry into the social security system.

Successful programs such as PLAN CEIBAL of Uruguay in the 2010s reduced the digital gap. Their free availability in the virtual world increased equal and unlimited access to information and improved e-government opportunities.

Dr Bosque (negative):

It seems to me that stability was the highest priority for decisionmakers. Traditional economic development decisions continued to strengthen the old production structures. Highly sophisticated consumerism came into the scene, but technological innovation was focused in the corporate sector, without real benefit for the larger majority of the population. The most advanced technologies were used only by the elite youth, while most people didn't understand them. Despite the creation of ministries of science and technology, the need for social development occurred not because of technology, but despite it. We needed public policies with "technological inclusion," assuring the ethically responsible use and democratic assessment of the benefits and potential hazards of emerging technologies like nano- and biotechnology.

The continual dumping of chemicals on the land and in the water and of carbon into the atmosphere is a byproduct of production. We all know the environmental and public health consequences. The online networks open the possibility for serious security flaws and terrorism for industrial operations and electronic transfer of funds. That some local companies joined the global campaign of corporate social responsibility is commendable but hardly persuasive. Environmental and social standards had to be imposed and enforced by legal mechanisms. Using environmentally friendly technologies is mandatory, not optional.

Dr. Arbusto (positive):

As a safety measure, auditor-directors elected by the community are serving on the boards of large companies, and purchases exceeding 1% of reported capital are made public through the companies' Web sites. Publicly owned entities (including universities) submit their balance sheets to private auditors. Mergers and acquisitions of domestic enterprises by large corporations are subject to legislative approval. By a law established for public credit, banks are required to allocate 50% of their loans to local investment and new business developments. A tax was introduced on loans to corporations for advertisement and promotion of innovation that is not in the public interest. The point was to encourage inventions and new products that brought public benefits with them while discouraging monopolistic incentives. A unique addition was the Open Innovation Forum, a think tank where participants openly discussed ideas that might lead— after more R&D—to patents.

The cyber revolution and participatory democracy expanded the e-government systems and made them more transparent. Most government bids are made via Web, and important decisions are referred to the people via e-referendum. Now, truly the people have a say on the solutions to everyday problems as well as major decisionmaking on long-term challenges. With online voting on new decisions and issues of general interest, recognition of new groups with common interests formed via Internet and social networks. Old problems still exist, but life is better for most people. Maybe things will be even better tomorrow, like the struggle for the elimination of food insecurity.

Dr Bosque (negative):

Although inequality has decreased, large migrations from the poor and disadvantaged regions to cities and developed countries are still occurring. Although living conditions and living standards are slowly but surely rising, there is still a significant mismatch between the technologies invented or acquired in Latin America and the needs to solve the region's massive social problems. Technology apparently has a mind of its own, evolving in ways that make attractive products and increased profits for large corporations but in its mindless way is not able to influence to any great extent the problems that plague our society; with some exceptions, of course…

Dr. Arbusto (positive):

Think of Latin America as a whole and not as a many different countries. Latin American integration exists in trade, as a beginning. We have institutions of continental cooperation to work for the people and biodiversity. Food security has improved, with pure meat produced without growing animals using cellular processes, and genetically modified foods have increased harvests.

And we are proud of our Latin American University, which enables the exchange of knowledge and ideas through student mobility.

We still have to find new ways to make the benefits of progress reach those social sectors most in need— including small local businesses. This would help ethical, social, and economic development and would reduce corruption, criminality, and social discrepancy.

Dr. Bosque (negative):

We can agree on that, at least if there are safeguards for the environment and ecosystems, natural resources, and native peoples' rights.

Scenario 3

Region in Flames: This report is *SECRET*

Date: July 31, 2030

This report is our manifesto for action and will be discussed at the committee meeting tonight. It is being sent to you in this way to avoid cyber-interception and blocking.

Yesterday, the last of the great independent newspapers in Latin America was burned to the ground. We all know it was not an unfortunate accident, as the government claims. There are now no newspapers or reliable news sources on the continent that can freely write their opinions about the true state of affairs, except the cyber underground and, as in this report, our private internal committees.

Here's what caused the latest attack on the freedom of the press. Their editorial, run on the front page, said:

The trends of the last two decades—drugs, corruption, poverty—have come together to create a situation that is worse than we could have imagined.

Families do not know where to take refuge. The drug chain has specialized by following the trends of legitimate business. Bolivia and Peru concentrated in production. Colombia and Mexico are carrying out the management—the intangible part of the business and the most profitable.

Bolivia, Colombia, and Peru have expanded coca cultivation. The cartels have taken over Brazil, Ecuador, Mexico, and Venezuela. These countries are living in a state of siege. The laboratories for processing coca are proliferating to other Latin American countries.

The fight amongst the drug heirs continues; it is an endless war. The United States of America is the main customer and a major financier. The drugs market is expanding with the production and consumption of synthetic opiates: the old ones such as amphetamines, ecstasy, and prescription drugs, and new ones that appear almost monthly. In Latin America, juvenile drug addiction is growing exponentially, mainly driven by marijuana and cocaine consumption.

The cartels murder migrants and kidnap people. The business includes trafficking of people and weapons, piracy, extortion, forgery, smuggling, predatory lending, and environmental degradation. A point of political instability has been reached, and governments are becoming increasingly corrupt. Money is not an issue and

honesty ultimately has a price.

The implementation of policies to improve employment, education, and social assistance systems has to be urgently fostered by UN agencies. Everyone seems to know that a coherent and sustained commitment is required from the U.S. government in coordination with Latin American governments, to combat international organized crime, but none of the so called "effective plans" have yet succeeded. Sustained demand, clever criminals, and endless bribery money have killed the most hopeful schemes. The rhetoric sounds great: "Latin American state policies have to be established against structural corruption and international drug trafficking with regional cooperation and stronger systems, creating an international fighting force," but so far, there are almost no lasting or effective accomplishments.

The legalization of drugs is still a matter of discussion. The World Health Organization says that it is not the answer, as it would create more health problems than solutions. Others think that legalization is the only way to dismantle the cartels and their systems of profits and power and to generate more tax money from drugs trade.

Education has been another promising possibility: teach the children as soon as they are old enough to understand that drugs are enslaving. Such plans are costly and, even in schools, money corrupts.

Meanwhile there is no agreement, no coordinated planning, no means to resist the flux of money, and the region lives in a brutal world where the worst is yet to come.

Our committee's research confirms this information. Freedom of expression on the continent is limited in several ways—some obvious, others not. In some countries, legislative and judicial powers are at the service of dictators. Political parties, unions, and all organizations are more or less controlled. All elections seem to be arranged and all candidates seem to have been chosen by the narco dealers. Most people are afraid to speak out. The people have traded democracy for survival.

The newspaper that burned down was not alone in its views. Last year's OECD report "The Outlook for Latin America—2030" noted that drug and corruption problems across the region have made it increasingly difficult to escape from the cycle of poverty. Poverty and misery have increased as the forms of crime have multiplied and the whole region has become "the kingdom of inequality." Income distribution in Latin America is the most unequal in the world, and GDP keeps falling. Most people are vulnerable, due to their lack of education, employment, and health services.

Projects for transforming public education—as a mechanism to equalize opportunities—have not succeeded yet. However, many access Internet tele-learning applications outside government authority. Clearly, realistic systems for allocating money to needy families with children are missing. Government-controlled education has poor coverage and poor quality. So the "no-no" (no study and no work) population continues to grow, more slowly than in the past but growing nevertheless. People have few opportunities for starting an independent life. The dynamics of inequality in education, income, and health reinforce each other in what seems to be an endless cycle and is further strengthened by lack of security and justice.

The magnitude of the problem of crime seems to be beyond the control of any agency acting alone because many—if not most—governments are corrupt and have tight links with international crime organizations. New "guerrillas" and terrorist groups are emerging. The old victims become criminals for surviving. There are no values. Getting involved in crime seems to be the only job option, the only viable way for surviving. Crime has become poor people's "modus Vivendi."

The *International Cyber-News* magazine, building on the OECD report, accused the international community of looking the other way and only becoming involved when the problem affects their own direct interests, such as cross-border violence entering the U.S.

Most Latin American governments have openly rejected the OECD report and the ICN story as being constructed by people who do not understand the region and profit from its decline.

With this background, it is understandable that the Economist Intelligence Unit reports that Latin America is poor and becoming poorer. GDP, exports, exchange rates, and remittances are all falling. Inflation is rising. In the region, economic uncertainty prevails and labor markets remain depressed. Governments are inclined toward fiscal austerity and reduced spending. Latin America is in recession and some would say depression, with problems of external debt, internal debt, and budget imbalance. The official unemployment rate is nearly 50%. Of course, many of the "unemployed" are really employed in crime and the drug industry.

The pace of growth in the region has declined due to lack of policies involving social, private, and public participation. The lack of confidence of foreign investors because of corruption and the vicious cycle of the lack of domestic investment in productive capacities is deteriorating the region's global competitiveness.

The Latin American countries continue to export raw materials and import finished products, hence making a net loss for the countries. There is a clear inability of the governments to centrally coordinate the resources and of the communities to self-organize and defend their own interests. And yet people argue that the market forces do their job with less government intervention.

The United Nations Environment Programme reviewed the situation recently and found that carbon dioxide emissions have tripled. Latin America and the Caribbean coasts are polluted by waste, garbage, chemicals, plastics, and especially fecal matter. They have lost much of their marine habitat, and human health is greatly affected. Hepatitis, cholera, diarrhea, malaria, dengue fever, and skin diseases are proliferating.

Coastal erosion, melting glaciers, forest fires, and flooding of beaches have driven away tourists. Deforestation, biodiversity loss, and lack of environmental governance are evident. The stability and productivity of ecosystems have been affected. Because of climate change, most countries in the region are vulnerable.

Droughts are frequent and some scientists have warned that the region may be on the threshold of a great famine. Human systems have become highly sensitive to changes such as water supply and demand, land use practices, and demographic changes.

The environment suffers from the "tragedy of the commons" in which no one is responsible. Polluters are not punished. There are no guarantees for foreign investment. Ecological tourism has vanished because biodiversity is disappearing.

Many human rights organizations, including Human Rights Watch, have condemned the state of affairs of most Latin American countries. They rightly say that we live in a region where poverty and youth are criminalized. Violence against women and children has increased exponentially. Torture and terror are common. Indigenous communities are increasingly homeless, marginalized, discriminated against, harassed, and abused by powerful corporations who want their land, mines, and forests. African-descendant communities and lesbian, gay, bisexual, and transgender people are living in constant danger.

Health care, clean water, education, and decent housing are not available to most people. The death penalty has been reinstated, and people live in an atmosphere of public insecurity. People from small socially excluded communities are despised, and young people are killed to collect the premiums paid for each "dead guerrilla." Activists and human rights defenders are constantly threatened, tortured, imprisoned, and/or killed. This is the decade of chaos, impunity, and dehumanization of humanity. Local gangs vie for power. Nobody knows where the newly disappeared have gone.

In Latin America and the Caribbean, submission, control, and authority of men over women persist. Gender-based violence includes sexual, economic, community, institutional, media, stereotyping, and biased ownership policies. Racism and discrimination persist. Maternal and infant mortality have increased. The law is not enforced, and religions condemn women for abortion. Women lack public health care, education, and opportunities to work. Mistreatment of women is increasing. Some policies that have been designed to correct these inequities have been attempted, but unfortunately they have not had much

effect. Police abuse continues. The Committee for Women's Defense has been created to force Latin American governments to dignify women's role and equity, but now the horizon is frightening.

The perception of what has happened to us is emerging in cyberspace and is being encouraged by committees like ours.

So, my friends, this is the background. The question is, Where do we go from here? Are we really inept? Are we really powerless? While the social revolutions against dictators and corrupt officials in Africa and the Middle East in 2011 were an important warning, political and drug leaders in Latin America didn't pay attention. The dictators are not attending to population needs and have succumbed to the pressure and money of criminal groups. Dictators feel themselves as Messianic, keeping their feudalistic patterns. They wish to avoid being judged by the International Criminal Court for theft and violating human rights. Corrupt officials profit at the expense of the population; their wealth grows while poverty and squalor persist. The drug leaders think they are invincible because they think they can buy anything.

Countries of the continent do not lack laws, constitutions, or systems of justice and redress. The framework is here. What has proved to be lacking is the will to enforce these structures, the ethics to do what's right.

The floor is open. Do I hear any suggestions for lawful action?

Scenario 4

The Network: Death and Rebirth

An ancient myth describes an attempt by the Babylonians to build a great tower that would reach to the heavens. The Babylonians believed that the invention of the brick—a real breakthrough at that time—would enable them to build such a structure. They were punished for their hubris; the curse of multiple languages destroyed their ability to communicate. In Latin America, we attempted to build such a "tower" and, as in the ancient myth, communications were our downfall—not words, but understanding of the meaning of words. Did we hope for too much? Our "tower" was a political, economic, and technological union of countries—a network called Cyber 1.0. Some people called it *Babble 1.0*.

This union attempted to integrate people from all Latin America and the Caribbean with different languages and from different fields of knowledge, with different values and ideologies. They were organized by thematic groups such as health, education, governance, environment, transport, entertainment, technology, conscience, citizenship, well-being, and happiness (rather than only economic success). It was a unique variation of the European Union. The members—individuals, groups, and countries—were called *Babelites*. The mission of the network was to advance political integration, avoid military conflicts, assure peace on the continent, prevent poverty, detect and reduce corruption, enhance economic development, improve decisionmaking, and foster social equity, as well as promote bottom-up development and empowerment. The Network provided a forum to enable fast and efficient interchange of ideas and information and improve participatory democracy. Idealists who were participants also hoped to change development paradigms while reducing rich-poor gaps, promote a worldwide friendship and fraternity without destroying cultural and natural diversity, return to traditional "indigenous" values of close communion with the environment, and increase the power of collective responsibility.

At the beginning, the network appeared to be the foundation of a continent-wide forum for participatory democracy, encouraging sustainable cities and consumer consciousness, citizenship, and a productive medium for social and environmental activists. Based on highly advanced communication devices and networks, the Cyber 1.0 platform was light-years ahead of the social networks of the last decades. It included multilingual with 3D holographic screens, global autonomous and seamless language translation, fast intelligent communicating engines, personal stories, socioeconomic data, happiness measures, and government goals, as well as votes on social and political issues in real time. It could recognize discussion themes and create ad hoc links to form automated topic forums, based on WSAI (for Wiki Semantic Artificial Intelligence) platforms. This forecasting capability stimulated the hope that the network would facilitate gradual Latin American integration and rescue the wisdom of its leaders and elder statesmen and women to build on traditions and to reinvigorate the "Latin Soul." However, as Colombian Nobel Laureate Gabriel Garcia Marquez implied in his *One Hundred Years of Solitude*, this movement was neither strong enough nor efficient enough to overcome the archaic institutional structures of the entrenched sociopolitical systems that were more or less in conflict with emerging social interests and values fostered by the embryonic union and its voice, Cyber 1.0. Ancestral cultures of the continent were swept away by the alienating consumer society. Therefore, like the Babylonians before them, the Latin American Babelites had goals beyond their capacity. And the story of collapse, like the fable of Babel itself, warned us about the ethical and economic dangers on the horizon for the sub-continent.

Outwardly, technology (like the Cyber 1.0 network) glittered and the economies of the sub-continent grew as a result of inventions, low labor costs, demand for its exports, and foreign investment. But poverty and inequality increased as well in the region. Why?, asked the people. Why?, asked the academics. Why?, asked the politicians. Why do we have apparent prosperity on the one hand and misery on the other, with high gaps in between? The collapse of Cyber 1.0 was a symptom of the ills that plagued the continent. Even though large corporations kept pushing development with an apparent technology boom and expansion of markets, high value-added products were increasingly imported and natural resources increasingly exported. The high rate of unemployment was a result of lack of social inclusion and accelerated automation of production processes; about 70% of the overall working population was at the margin of the economy.

The media were still controlled by one big group called EPIC: Established Politics for Information and Communication. Using all channels of media, TV, digital, printing, and so on, the media monopoly was in charge of what many critics liked to call BBS2, for Big Brother and Sister 2—in fact, it was "transliminal mind control" or at the very least manipulation of the population based on bread and circuses. Like a new generation of Wikileaks, Cyber 1.0 was the enemy of EPIC, and some people believe the network's disintegration came in part from EPIC's "dirty tricks," which included implanted misinformation and even using Radio-Frequency Identification chips on people. The tragedy is that nobody seemed to care that the network that had a chance of reforming the politics and economy of the region had passed from the scene. What emerged instead was a new form of collective manipulation by powerful elite. Massive state and economic control slowly destroyed cultural differences, creating homogeneous thinking and a continued disintegration of the common social threads holding society together. EPIC encouraged consumer aspirations and "mass mind" ideals, starting from an early age. Australian political analyst Sharon Beder wrote about this possibility a long time ago in her classic work regarding the corporate capture of childhood. The faith that things can be achieved through cooperation was lost.

While the political hopes for continental unity and communications were fading, the world of Latin America was changing. The warnings from the Rio+20 Summit in 2012, where experts agreed that weather phenomena would have greater intensity, were confirmed. As a matter of fact, developing regions like Latin America have proved to be the most vulnerable to climate change and variability. Social unrest grew, partially as a consequence of climate change, which reduced water from the Andes (thus lowering rivers' debit), changed insects' migrations, and altered human disease patterns. These were aggravated by intensified rainforest deterioration, crop failures, uncontrolled use of GMOs, and destruction of regions important to indigenous tribes (that started with the Brazilian Belo Monte dam in 2011). Migrations have doubled since 2011, mainly due to water and food shortages.

Daniel Pérez, an Argentinean environmentalist from the Cyber 1.0 era who attended the Rio+20 Summit, declared that Latin America is a continental environmental failure, noting that his country now uses transgenic seeds for 100% of the crops instead of adopting a green business model in its agriculture. Even worse, Brazil continues to be the largest consumer of pesticides in the world, compromising its enormous natural biodiversity and natural water resources so vital to the rest of humanity. Other activities with serious environmental consequences across the continent included political corruption leading to privatization of water reserves, lack of effective waste management policies and sustainable consumption and production approaches, radical changes in the courses of rivers, poorly controlled use of nuclear energy even after the 2011 Fukushima warnings, and mining exploitation without ecosystem restoration. The obsession for producing biofuels and the expansion of coca plantations affecting biodiversity took priority over sustainable food cultivation.

When the network failed, political unity among nations fragmented as well, and civil apathy in Latin America grew. The rule of law gradually lost its grasp, allowing organized crime to take over states with increasingly corrupt governments. Countries like Colombia, Venezuela, Bolivia, Paraguay, and Mexico slipped completely under the control of organized crime groups. Rio de Janeiro had a semi-legitimate "narcodemocracy," where people's representatives met weekly with cartel gangs to monitor the action-plan established by the Narco Vision 2035 project of the Latin American Narcopower Cartel, launched in Favela Rocinha around Rio de Janeiro in 2015.

However, there is a glimmer of hope. Quietly, Cyber 2.0 has taken shape; it features all of the interconnectedness of the Cyber 1.0 network but it now includes truth checking (displays the likelihood of a statement being true and authentic)

and increased government participation, which appears to be fostering a spirit of multilateralism once again. The network also includes detectors of corruption and advanced anti-virus detection and prevention, while governments have enacted new laws that make tampering with the network a crime. Some people called this new rebirth *Agora*, after the Greek forum—a place of free speech, trade, and open political debate. In addition to curbing the power of organized crime, the achievements of this new network of like-minded reformers already include improved ability to foster true dialogue and empower communities to devise and implement solutions, as well as tools to help decisionmakers in all sectors together with government—instead of by government—to improve the quality of their decisions, thus increasing the strength of civil society and promoting social integration. Projects for improving indigenous peoples' rights to resources and new community-oriented development are devised cooperatively online. Online social movements and e-government systems help reduce corruption and improve decisions. Internet access is free, and knowledge is shared through the implementation of UREDAP, the Urgent Educational Attention Program, a network to expand global knowledge to the excluded segments of population and all areas of society through local cyber centers.

In this chaotic time of corruption, exclusion, cultural fragmentation, ecological damage, and strong cartels for information, energy, and raw materials, another Latin America is becoming possible: Cyber 2.0 led the political arguments for open-sourcing of green technologies for the development of integrated smart and clean energy distribution systems everywhere. Energy systems now permit houses to receive and provide energy to and from a parallel new energy distribution network. Such energy enterprises were described and encouraged on Cyber 2.0 and garnered the support of large investments from big pension funds and venture capitalists from all over the world.

In particular, new advanced intelligent energy grid systems involving small farmers and communities earn income from energy production while also protecting the environment. The region's biodiversity became essential not only for the planet's health but also for assuring the availability and abundance of organic foods, cosmetics, and biopharmaceutical products. Other benefits of the new energy strategy were more than self-sufficiency, the exportation of energy surplus to other low-income countries with less capacity, and new opportunities for small businesses in the neighborhood, allowing a massive contribution to the local economy. These leapfrog innovations have produced positive impacts: better quality of life, reduction by 75% of the production of fossil fuels, and improved energy security. Communities now can decide what types of energy they wish to use, with benefits and downsides proper to each community's main industries, using the supplement of massive energy production by large power stations necessary for large populations and small-scale energy production suitable for small and remote communities. The savings resulting from the optimization of these new systems have been allocated to education, housing, and health and social care programs. The investments in local economies have reduced the migration from rural to urban areas, and more and more people are willing to go back to the quietness and healthiness of rural life.

Foreign investment has been high, and despite increased crime, tourism is up and Latin America now competes with Europe as a travel destination. Yet inflation is always a great concern and could again affect development. Throughout the last decade, the idea of a common Latin American Peso and a complete geo-political and economic union with coordinated standards and trade agreements was continually discussed. But the process may take even longer than the creation of the European Union because of the internal interests of the member states and external pressures. Instead, Cyber 2.0 has led to an era of multilateralism—agreements among countries that see it to their benefit to agree on standards, laws, and behavior norms.

After the last massive financial crisis at the beginning of the century, leading economies of Latin America have increased their power within the G-40—now led by China and India—even if the world still lacks good economic policy coordination. While the European Union is still pushing the social union approach, Latin America only cares about the economic one. Therefore, the LA Union is now committed to developing new finance and economic regional institutions to foster sustainable development, by institutionalizing common economic policies.

The death of Cyber 1.0 and early plans for continental integration set the stage for the rebirth of both. Today our continent has the best of the world's science and technology: whatever is available in the world is available here, but maybe not to everyone. Technology brings the allure of materialism. We love it, we think we need it. We have the highest computer speeds and the smallest chips; Cyber 2.0 is the least costly and highest speed network anywhere in the world. But has technology helped us, really? We still have poverty and crime and corruption, battered women and starving children, homelessness. We have the worst the world has to offer, too. We can modify DNA and produce marvelous new life forms that detect and cure disease and shape our prospective progeny, but also the growing spectrum of bio-weapons and grave uncertainty about the morality of using technology to change human destiny and evolution. We have nanotechnology and the worst slums in the world; we grow sugar for fuel when we need food. We make decisions that affect the world without knowing the consequences, using obsolete and impoverished decisionmaking.

The scale that balances technology against social need is tilted strongly toward technology. We haven't yet learned how to use it to make Latin America work for the best of all, but we may be learning. We need to redesign government policy and legal standards directed to making technology breakthroughs and applying knowledge for the benefit of all and of nature. New actors that were before excluded in important decisions are now on board in this movement: women who are key players in the economy and education, youngsters who are now pioneering with social enterprises, and elderly people whose population has doubled since 2011 and who are now engaged and feeling the call of the earth too. Our network of bilateral agreements slowly spreads these goals among Latin nations.

By this time, in the beginning of the 2030s, a new decade is foreseen for Latin America's future. Change is already overdue and is more than necessary for the survival of the whole continent, envisioned as a promising Up from Eden yet-to-be accomplished process, but now more aware of who and where we are, and where we want to go. For that reason, dystopias are being replaced by utopias, and a whole new set of AQAL (for All Quadrant All Levels) Well-Being Indicators is emerging.

Cyber 2.0 is a step toward improved democracy and social evolution building on the cultural roots of our continent. The whole continent is now on the move, in search of new leadership models, cultural identity, regional integration, ethical values through education and culture, and solid leadership fundamentals to promote solidarity among nations—rethinking the real purpose of knowledge and heading for new solutions. A rebirth is on the way. But it may depend on, as in many other regions, the extent to which we may go regarding a new framework of a trusting and empathic civilization.

Conclusions

The previous scenarios can be used as input to national and regional policy planning processes. Decisionmakers can use these scenarios to ask how their policies might fare in the range of situations depicted by these scenarios and find courses of actions that produce desired results in all the situations depicted by the scenarios. It is helpful that a set of alternative scenarios illustrate a reasonable span of plausible futures and that the content in each is detailed enough to reveal potential impediments and opportunities for action.

There is less consensus about the development model in Latin America than in more developed countries. Basic forms of social organization, such as markets, states, and institutions, or the separation between society and state are not universally accepted in Latin America. The conflicts within these countries (ethnic, cultural, economic, social, political, and technological) have not been resolved. Tensions between modernization, development, and the "traditional" social and economic organization are most evident in a context of rapid changes in technology and globalization.

As these four scenarios suggest, Latin America must find its place in a changing world while it simultaneously meets the basic and expanding requirements of its population in education, health, housing, jobs, safety, and other services for individual and social development.

If deeper integration of the region is to be achieved, common strategies and policies must be sought. An improved economic framework would include matching national production systems and markets across national boundaries. With proliferation of the Internet, integration of educational systems in the region is possible, as is the meshing of science and technology systems. Local and regional innovation policies are needed. The informal economy should be integrated with the formal economy, while establishing a new balance between external and domestic markets.

The biggest challenges are to recognize and incorporate the requirements of future generations in the formation of public policies and to create a balance of opportunity costs for human, natural, and technological resources in each country. Solutions must be found to problems that reach across the scenarios: corruption, crime, and drugs. Until progress on these is achieved, the wisest of development strategies is not likely to succeed.

As imaginative as these scenarios may be, they certainly omit surprises that may lead to disruptions in society, infrastructure, businesses, and economies. The key to effective response to such exigencies is resiliency.

After 200 years of Latin American independence, a reorganization is evolving as a result of internal pressures and globalization. New relatively short-lived initiatives have come and gone because of an essential lack of real Latin American identity, on the one hand, and globalization of multinational lobbies fighting for their own interests on the other hand.

The scenarios include many examples of positive initiatives related to high-tech social networking: Cyber 2.0, 3D holographic screens, seamless language translation, fast and intelligent communicating engines, expanding socioeconomic data, happiness measures, WSAI platforms, smart-grid energy systems, the exchange of knowledge and ideas through student mobility, cyber revolution and participatory democracy, e-government systems, new soft technology to improve decisions, transnational power blocs to promote a more sustainable world, construction of a Latin America University, consideration of a Latin American currency, a 2030 Mexico City World Expo consolidating the futuristic image of Latin America, "Made in Latin America" brands revolutionizing international trade, and so on.

All four scenarios are powerful resources in understanding the threats and opportunities of the future. The rebirth of Latin America may be on the way, but this rebirth may depend on how much and how fast we move toward new frameworks of institutional power and new paradigms in governance. We also need to transform educational systems to develop a new generation of leaders who cultivate and share ethical principles in their decisionmaking, understanding that the ultimate meaning of life is to expand human potential and well-being in such a way that the next two decades may be promising in terms of a better Latin America for a better world.

Key environmental and resource constraints, including health risks, climate change, water scarcity and increasing energy needs will further shape the future security environment in areas of concern to NATO and have the potential to significantly affect NATO planning and operations.

Active Engagement, Modern Defence. Strategic Concept for the Defence and Security of the Members of the North Atlantic Treaty Organization adopted by Heads of State and Government in Lisbon, 19 November 2010

While climate change alone does not cause conflict, it may act as an accelerant of instability or conflict, placing a burden to respond on civilian institutions and militaries around the world.

2010 Quadrennial Defense Review
United States Department of Defense

6

Emerging Environmental Security Issues

Environmental security is increasingly dominating national and international agendas, shifting defense and geopolitical paradigms. Climate change and unconventional security issues—impervious to national sovereignty, ideology, and military power—are now recognized as top threats to peace, political stability, and prosperity. The role of environmental diplomacy is growing and environmental security–related concerns are becoming defining factors in international political and military negotiations.

The dynamics of security strategies are changed by the new circumstances and forecasts, demanding cooperation on non-traditional threats such as natural disasters; potential biological, nuclear, or chemical terror; water, food, and energy security; and increasing environmental and social problems, as well as the deepening gap between those who could cope with the effects of climate change and those who could not. These challenges are so complex and changing so fast that it is increasingly difficult to design realistic long-term strategies and impossible for any single nation to address them alone.

International law systems and organizations are adjusting to better support environmental security, from the protection and management of natural resources to liability for environmental damages. The ability to identify environmental threats and crimes has been strengthened by increasingly powerful detection and monitoring technologies and by environmental jurisprudence supported by improved enforcement mechanisms. Environmental damages that people and organizations got away with in the past are less likely to escape exposure and punishment in the future.

The Millennium Project defines environmental security as environmental viability for life support, with three sub-elements:

- preventing or repairing military damage to the environment
- preventing or responding to environmentally caused conflicts
- protecting the environment due to its inherent moral value.

This chapter presents a summary of recent events and emerging environmental security–related issues organized around this definition. Over the past several years, with support from the U.S. Army Environmental Policy Institute, The Millennium Project has been scanning a variety of sources to produce monthly reports on emerging environmental issues with potential security or treaty implications.

More than 300 items have been identified during the past year and over 2,500 items since this work began in August 2002. The full text of the items and their sources as well as other Millennium Project studies related to environmental security are included in Chapter 9 on the CD and are available on The Millennium Project's Web site, www.millennium-project.org.

Preventing or Repairing Military Damage to the Environment

Since conflict and environmental degradation exacerbate each other, their spectrum and severity could expand unless they are addressed together. Defense experts increasingly argue that environmental security should be on a par with and an integral part of conventional security. Including environmental factors in military actions gives strategic advantages in combat and post-conflict operations, protects the health, safety, and security of the troops, and develops diplomatic relations and the confidence of local populations and neighboring countries, thus increasing any mission's success.

NATO's Strategic Concept for the next decade stipulates that the world's security environment and the organization's planning and operations will be increasingly shaped by key environmental and resource challenges such as climate change, water and food scarcity, and growing energy needs. The roadmap has been also updated to consider modern threats such as energy security, cyber attacks, and the security impacts of emerging technologies, along with and in the context of the spread of terrorism and extremist groups.

The U.S. Army's Strategy for the Environment and new special projects show military leadership in protecting the environment, increasing energy efficiency through procurement and operations, R&D centers of excellence, and the transfer of knowledge. The *National Security Implications of Climate Change for U.S. Naval Forces* report by the National Research Council argues that climate change raises challenges to America's current naval capabilities, requiring serious changes to the design of their fleets, training, and ships' deployment.

Environmental factors are affecting both resource-scarce and resource-abundant countries. The most critical situation is in 40 or so fragile and conflict-affected states, where a growing young population compounded with scarce resources and unstable political systems deteriorate environmental security, further aggravating the vulnerability to violence.

The Global Peace Index 2011 shows that the world's peacefulness decreased for the third year in a row, mostly due to internal unrests rather than warfare between countries; the likelihood of terrorist attacks increased in 29 of 153 countries, while violent demonstrations increased in 33 countries. The cost of violence to the global economy was estimated to be over $8.12 trillion in 2010. USAID notes that over the past 60 years, at least 40% of all interstate conflicts had a link to natural resources. UNEP reports that since the mid-twentieth century more than 90% of major armed conflicts took place in countries that contained biodiversity hotspots and over 80% occurred directly within a hotspot area, further threatening biodiversity. The Pacific Institute's Water Conflict Chronology Map identifies more than 100 conflicts over the past 20 years that were water-related. While conflicts involving natural resources are twice as likely to relapse in the five years following a peace agreement, UNEP notes that fewer than 25% of relevant peace agreements address environmental or resource aspects. The 2010 Environmental Performance Index reveals that most lower-ranked nations are also vulnerable states.

The UN Security Council's focus on the environment-security-development nexus is increasing, as several countries are urging that climate change be addressed as a global security threat, with issues ranging from loss of livelihoods and illegal exploitation of minerals to the impacts of climate change on national sovereignty.

The UN Convention to Combat Desertification suggests adopting the concept "securitize the ground" in order to create a wider global political awareness of the social, environmental, and economic consequences of desertification, land degradation, and drought.

Lawyers and human rights activists are assessing legal instruments for prosecuting the pillage of natural resources as a war crime. While

this would primarily apply to companies profiting from the trade of "conflict minerals" and to cases that use resulting revenue to fund armed conflict, concerns also include environmental degradation and social aspects. The most notorious situation is the Democratic Republic of the Congo, but other countries on the "watch list" include Brazil, China, India, Mexico, and Turkey. The U.S. Dodd-Frank Act (H.R. 4173) that became effective in April 2011 includes a clause requiring companies to report on their use of certain minerals from the DRC and neighboring countries, with non-compliance being fined.

In the eastern DRC, illegal timber logging and rare mineral extraction have historically fueled conflict. UN aid workers estimate that 890,000 people are internally displaced in the provinces, while security forces in the region have difficulty maintaining their peacekeeping mandate, which includes the protection of civilians and, by extension, control of natural resources through training and other military assistance to the government.

Defense officials in developing countries increasingly see security in terms of food and water security and natural disasters. Often, there might be a dilemma of allocation of forces and funds between traditional and environmental security. In 2010, Pakistan's defense budget rose by about 17%, to $5.2 billion, while the July 2010 flooding that affected one-fifth of the country's land and about 20 million people, with a death toll of close to 2,000 and total economic loss of $43 billion, arguably had a higher impact than anything the Taliban could accomplish.

Measuring Progress in Conflict Environments, a project developed by a consortium of organizations working in development, security, and policy, provides a framework for analyzing the peace progress during stabilization and reconstruction in order to identify the drivers that impede or facilitate the end of conflict. The system was tested in Afghanistan and Sudan and is currently being applied to crisis cases for further improvements.

Although Protocol 1 of the Geneva Conventions contains text protecting the natural environment, UNEP notes that there are no mechanisms in place to protect natural resources during armed conflict and no permanent international authority to monitor violations and to address liability and redress claims for environmental damage in those situations.

The Convention on Cluster Munitions entered into force in August 2010, two years after its adoption. It bans the use, production, and transfer of cluster munitions and sets deadlines for stockpile destruction and clearance of contaminated land, as well as prescribing responsibilities toward affected communities. As of mid-2011, a total of 57 countries had ratified and 108 had signed the convention. This sets a precedent on how a "coalition of the willing" can successfully lead to international regulations, and it might trigger similar negotiations and be emulated for other weapons.

The first Review Conference on the Rome Statute of the International Criminal Court added the criminalization of the use of certain weapons in non-international conflicts under Article 8 (paragraph 2, e) and includes poison, poisoned weapons, asphyxiating, poisonous or other gases and all analogous liquids, materials, or devices, as well as the use of bullets that expand or flatten in the body. It also reached agreement on the definition of the crime of aggression and the framework for the Court's jurisdiction over this type of crime.

UNEP recommends that the Permanent Court of Arbitration and its "Optional Rules for Conciliation of Disputes Relating to the Environment and/or Natural Resources" should be considered for addressing disputes related to environmental damage during armed conflict and that a summary report on the environmental impacts of armed conflicts should be presented annually to the UN General Assembly.

INTERPOL's 79th General Assembly resolution recommends that the world police organization form an Environmental Crime Committee. It underlines that since environmental crime is multidisciplinary in nature and not restricted by borders, it has to be addressed at the global level, with INTERPOL and the National Central Bureaus playing a leading role. INTERPOL also created a Radiological and Nuclear Terrorism Prevention Unit for expanding its current anti-bioterrorism activities to address chemical, biological, radiological, and nuclear threats.

Environmental degradation and hazardous ordnance leftovers in many post-conflict areas around the world threaten the livelihoods and health of current and future generations and may constitute an impediment for lasting peace. Leaking abandoned ordnance since World War II and dangerously high levels of heavy metals and other toxic chemicals related to military exercises are contaminating the oceans, endangering the marine ecosystem and human health.

The war in Libya makes it impossible for that country to meet the deadlines of May 2011 to destroy its cache of mustard gas and December 31 to eliminate its precursor agents, as requested by the Chemical Weapons Convention. Japan's nuclear and environmental disasters might further delay efforts to complete its obligations to dispose of the chemical munitions in China. The U.S. and Russia are also unlikely to meet the 2012 deadline for eliminating their respective stockpiles of chemical warfare materials. As of end of April 2011, the U.S. had destroyed about 86% of the warfare agents it held when the treaty entered into force in 1997, while Russia had destroyed about 49% of its stockpile as of February 2011, according to authoritative sources.

INTERPOL's Project Geiger database launched in 2005, developed in collaboration with the IAEA and other organizations, lists over 2,500 incidents linked to illegal radiological and nuclear trafficking. Meanwhile, with the entry into force of the Pelindaba Treaty for an African Nuclear-Weapon-Free Zone, nuclear weapons are being banned throughout the entire southern hemisphere. The new Strategic Arms Reduction Treaty was signed by the U.S. and Russia (together holding more than 90% of the world's nuclear weapons), requiring each to reduce their strategic nuclear arsenal, although critics note that the treaty does not address the disposal of the nuclear material contained in the weapons. The UN Security Council resolution aiming to advance global nuclear disarmament stipulates that non-compliance with the Nuclear Nonproliferation Treaty would be referred directly to the Security Council rather than to the IAEA.

While the chemical and nuclear weapons conventions have enforcement mechanisms, the Biological Weapons Convention does not, but negotiations continue. Meantime, the threats of bio-error and bio-terror increase. Developments in synthetic biology, cognitive science, nanotechnology, electromagnetic pulses, and other high-tech fields, combined with the availability of information and low-cost components needed to produce WMDs as well as the increase of terrorism and social unrest (often exacerbated by environmental factors), are increasing the threat of terrorism and single individuals who could use bioweapons to be massively destructive.

After land, sea, air, and space, cyberspace became the "fifth battlespace" on the agenda of security experts. Disruptions of critical infrastructure, such as water or electricity by cyberattacks in an IT-dependent world, call for new legal and policy frameworks. Cybersecurity challenges include cybercrime, cyberespionage and reconnaissance, and cyber-leveraged and information warfare. The EU will create a new cyber-defense unit that will pull together IT departments from the European Commission, Parliament, and Council to share intelligence and address attacks on all EU bodies. The U.S. has released its plan to protect the nation's cyber infrastructure, while the Pentagon's planned new strategy could qualify a cyber-attack from a foreign nation as an act of war that may result in military retaliation.

New technologies are offering unprecedented detection, cleanup, monitoring, and surveillance possibilities for environmental security. Intelligent battlefield robots will have elements of the rules of engagement and the Geneva Convention built into their programming. A NASA project tested the concept of "spiderbots" that can be placed into a hazardous environment to communicate among themselves and with the outside world, including satellites, to monitor an environmental situation. Ultra-sensitive portable chemical and biological devices offer increasing accuracy in detection, monitoring, and cleanup, with rapid response time.

Preventing or Responding to Environmentally Caused Conflicts

The UN identifies five channels through which climate change can have security implications: impacts on livelihoods and vulnerable people, economic development, population migration and/or conflict over scarce resources, displacement of whole communities due to sea level rise and consequent statelessness, and access to internationally shared resources.

The WMO notes that 2001–10 was the warmest decade on record, with the global average temperature 0.46°C above the 1961–90 average. Meteorological organizations forecast that the intensity and frequency of extreme weather events will grow worldwide, and climate patterns are changing. By mid-2011, there were over 1,000 confirmed tornadoes in the U.S., causing an estimated 523 deaths (almost as much as the total for the previous 10 years).

According to the Centre for Research on the Epidemiology of Disasters, in 2010 there were 373 disasters registered, affecting 207 million people—89% out of whom were in Asia. In 2011, disasters had already caused more than $300 billion in losses by May, almost the same as in all of 2010. The UN estimates that the amount of global wealth exposed to natural disasters risk has nearly tripled from $525.7 billion 40 years ago to $1.58 trillion today. The risk of economic losses in OECD countries due to floods has increased by 160% and for tropical cyclones by 262% over the past 30 years. Calling for improved adaptation policies and funding, officials forecast that for every $1 invested in resilience and prevention, $4–7 are saved in response.

A Humanitarian Emergency Response Review estimates that around 375 million people will be affected by climate-related disasters every year by 2015 and many more by other "rapid onset" emergencies and the impact of conflicts. Climate Risk Index 2011 by Germanwatch shows that developing countries are among the nations most affected by extreme weather.

The Social Conflict in Africa Database includes over 6,300 social conflict events for the period 1990–2009. The pattern reveals more social conflicts in years that were extremely wet or dry than in years of normal rainfall. The Food Security Risk Index 2010 reveals that the countries most at risk from shocks to food supplies are also among the countries with serious security problems. Rated at most "extreme risk" are Afghanistan, DRC, Burundi, Eritrea, Sudan, Ethiopia, Angola, Liberia, Chad, and Zimbabwe.

Turning around the increases in world food prices will become increasingly important for stability. World food prices have more than doubled since 1990. Oxfam predicts that the average cost of key crops could further increase by 120–180% by 2030. This may understate the severity, since 16 factors that directly or indirectly increase food prices all look like they will be rising: population growth, rising affluence especially in India and China, diversion of corn for biofuels, soil erosion, aquifer depletion, the loss of cropland, falling water tables and water pollution, increasing fertilizer costs (rising prices for oil, phosphorus, and nitrogen), market speculation, diversion of water from rural to urban, increasing meat consumption, global food reserves at 25-year lows, climate change that increases droughts and desertification (in dry areas), flooding (in wet areas), melting mountain glaciers that reduce water flows, and eventually saltwater invading croplands.

While genetically engineered seeds adapted to a harsher climate could help increase yields, researchers warn that increasing corporate control over seeds is reducing the diversity of traditional seed varieties and traits that help farmers adapt to the effects of climate change, jeopardizing poor farmers' livelihoods and strongly influencing food prices.

The World Bank estimates that up to 30 million hectares (74 million acres) of farmland are lost each year due to severe degradation, conversion to industrial use, and urbanization. Additionally, large-scale land acquisitions in regions that are already food- and water-scarce, as well as the allocation of land to produce agrofuels rather than food, risk increasing poverty and social unrest. Within Africa's Sahel, a region of approximately 60 million inhabitants, extreme drought and unpredictable weather patterns continue to worsen food and water security and interregional migration.

If current trends continue, most glaciers in the mountains of tropical Africa will disappear by 2030, and those in the Pyrenees will be gone by 2050. Since 70% of fresh water is trapped in glaciers, once they are gone the situation for human survival will become critical. The *2011 Water Stress Index* reinforces that Africa and the

Middle East, especially countries on the Persian-Arabian Gulf, are most vulnerable to serious water shortages, increasing the likelihood of resource-based conflicts in these areas. The UN estimates that 18 of the 30 water-scarce nations will be in the Middle East and North Africa by 2025. The capital of Yemen is expected to run out of water much sooner.

Food and water issues are also considered to be exacerbating factors in the 2011 Arab Spring uprisings. The political turmoil could further affect living standards in the region, fueling tension in an already conflict-prone region. As the scope and spectrum of the protests expanded, energy security concerns around the world increased. Fear of extended interruptions in oil supplies from these countries rapidly drove up prices. Unreliable production and exports of oil from the region would cause greater demand on oil supplies from the North Sea and Africa.

The scale of the Japanese disasters (in a relatively well-prepared country) and the potential increase in the number and intensity of natural disasters around the world due to climate change trigger important reexaminations regarding preparedness and resilience, as well as the management of nuclear and other hazardous material. Political leaders are calling for a review of the IAEA's nuclear safety convention and for efforts to make the standards mandatory and enforceable, while restricting reactor construction in earthquake-prone areas. Many nations are changing their nuclear policies, with Germany and Switzerland now planning to completely phase out nuclear power.

Russia is building "ecological barriers" on its borders to reduce the impacts of future international disasters such as the oil spill in the Gulf of Mexico and the Fukushima nuclear disaster. A sensor network will monitor air and water pollution on the Russian borders, thus giving early warning of danger.

The annual demand for rare earth elements has skyrocketed over the last decade from 40,000 tons to 120,000 tons, and by 2014 this might increase to 200,000 tons, assuming green and IT technologies continue to increase. China, which controls over 90% of known rare earth supplies, has been gradually reducing export quotas since 2005 and might completely stop exports by 2012.

Disputes over deep-water oil territorial claims in the South China Sea and the Arctic are potential areas for conflict. The Arctic is warming faster than forecast, and human activities—from navigation to exploitation of natural resources—are increasing. The Seventh Ministerial Meeting of the Arctic Council, in May 2011, adopted the Agreement on Cooperation on Aeronautical and Maritime Search and Rescue in the Arctic, the first legally binding agreement negotiated by the Council.

A report by the Arctic Monitoring and Assessment Programme predicts that by 2100 sea level could rise 0.9–1.6 meters, depending on the rate of melting of the Arctic and Greenland's ice sheets, while new research found that ice loss from Antarctica and Greenland has accelerated over the last 20 years and is occurring faster than models predicted. This puts in danger the very existence of small island states such as Kiribati, the Marshall Islands, and Tuvalu in the Pacific and the Maldives and Seychelles in the Indian Ocean. The President of Kiribati says that in the country's outer islands the situation is already critical, as an increasing number of coastal villagers need to be relocated because of rising sea levels.

Experts are assessing existing formal and informal rules that would apply to shifting maritime baselines due to climate change. Such situations range from delimitation of maritime economic exploitation zones to the continued existence of some nations as legal and sovereign entities even if their entire population was forced to relocate elsewhere. Some potential options are updating UNCLOS with a concept of moving maritime baselines or making today's baselines and boundaries of maritime zones permanent.

WMO is developing the concept of hydrometeorological security with a global framework for climate services for better integration of global observing, information systems, and disaster risk reduction.

Protecting the Environment Due to Its Inherent Moral Value

UNEP asserts that an investment of 2% of global GDP per year in 10 key sectors could trigger "greener, smarter growth," removing the inherent risks and crises associated with the current "brown economy" model, while investing about 1.25% of global GDP per year in energy efficiency and renewable energies could cut global primary energy demand by 9% in 2020 and close to 40% by 2050.

While there is general agreement that there are gaps in the current environmental governance system, views differ about potential solutions. Some countries favor creating a global policy organization with universal membership to manage the global environmental agenda, while others advocate a new specialized UN agency on the environment or argue for an umbrella organization on sustainability. However, there is general support for other broad reforms, such as setting up an all-encompassing global information network, establishing a tracking system on environmental finance, and enhancing UNEP presence within existing UN country offices.

There are more than 700 multilateral environmental agreements, and the focus of international negotiations is switching from designing new treaties to reinforcing existing ones and strengthening international environmental governance. Following the successful synergies developed among the three conventions on chemicals and waste—the Basel, Rotterdam, and Stockholm Conventions—a framework for coordination of all biodiversity-related MEAs and UN bodies is being created. Considering impediments, six conventions form a potentially manageable and coherent cluster: CBD, CITES, CMS, Ramsar, WHC, and ITPGRFA, while the CBD, UNFCCC, and UNCCD cluster would assure a better integration of biodiversity with climate change issues.

These synchronizations would improve global environmental governance by increasing coherence in decisionmaking and monitoring at international, regional, and national levels. Integration is also being initiated among regional regulations. For example, China, Japan, and South Korea have set a broad framework for adapting their chemical regulatory systems to the EU Registration, Evaluation, Authorization and Restriction of Chemicals (or REACH) system, and in May 2011 they decided to foster cooperation on non-traditional threats such as nuclear safety,

disaster prevention, and food, energy, and environmental security.

Evaluation mechanisms of the effectiveness of agreements are improving, and increasingly powerful analytical tools are being created to compare national environmental status. New international watchdog bodies have emerged, and others are being proposed to assist legal action against environmental crimes. And indexes are being created to measure progress and assess policy efficiency or to set priorities.

There is a growing trend for an ecological democracy, with people demanding active participation in decisions that have ecological impact. The Protocol on Strategic Environmental Assessment to the UNECE Espoo Convention sets the legal framework for better integration of environmental and health assessments, as well as public participation in decisionmaking at the earliest stage of projects and programs. The Lima Declaration on mining calls on governments to enact measures limiting (or revoking) transnational companies' rights to mine on indigenous land without previous consultation with the indigenous people. It calls on the UN to declare indigenous peoples "the rightful owners since the ancient times of the soil, subsoil and natural resources" of their territories, and also attests indigenous people are "committed to instrumentalize the International Court of Justice Climate" and the "construction of a national and regional agenda for climate justice."

Bolivia is preparing a draft UN treaty on the Rights of Mother Earth, similar to that on human rights. The treaty aims to institute 11 rights protecting nature from human intervention, ranging from the right to clean water and air to unaltered vital cycles and equilibrium and the right to not be genetically modified. It builds on President Evo Morales's proposal in January 2010 for an international court for environmental crimes and the "Rights of Mother Earth," as well as a Bolivia-led UN resolution in 2009 that proclaimed April 22nd International Mother Earth Day.

An International Consortium on Combating Wildlife Crime was formed by INTERPOL, CITES, UNODC, the World Bank, and the World Customs Organization. It will improve coordination of the five organizations' work to curb wildlife crime, which is generally transnational and involves several types of criminal organizations.

The 2011–20 Strategic Plan for Biodiversity identifies 20 targets, including expanding the world's protected areas to include 17% of terrestrial surface and 10% of the marine surface; the restorating of a minimum 15% of ecosystems already degraded; and halving, or bringing as close as possible to zero, the rate of loss of the world's natural habitats. Supplementary new protocols to the CBD provide international rules and procedures for liability and redress related to living modified organisms, geoengineering, and use of genetic resources.

Although global agreement for a post-Kyoto treaty has not been achieved, more local progress is being made. China is drafting a national law dedicated to climate change and its latest Five-Year Plan (2011–15) is switching its focus from GDP quantity to sustainable quality. After the EC plan to increase energy efficiency by 20%, increase renewable energies to 20%, and reduce greenhouse gas emissions by 20% (20/20/20 plan), the Community put forward a "roadmap" for achieving a low-carbon economy by 2050 in the EU. The EU is also considering creating a "coalition of the willing" to continue the fight to reduce GHG emissions in the absence of an international treaty.

The emergence of nanotechnology and synthetic biology and the proliferation of personal electronics bring new international environmental security requirements. The International Organization for Standardization publishes a globally harmonized methodology for classifying nanomaterials. The EU is exploring how to include nanomaterials within the REACH context.

Many research activities around the world are evaluating health and environmental implications of nanotechnology as well as mechanisms to reduce their negative impacts.

Since synthetic biology could one day be misused to create bioweapons and potentially even WMDs, international agreements to regulate this new technology seem both likely and warranted. The scale and scope of the expected future biological revolutions may one day require an international regulatory agency similar to the International Atomic Energy Agency.

Electronic waste is an extremely serious problem that is not getting adequate attention. It grows by 40 million tons a year around the world, and it is expected to rise dramatically in developing countries, which are vulnerable to illegal trafficking of hazardous waste unless regulations are strengthened and enforced. Computer waste in India alone is projected to grow by 500% by 2020 compared with 2007 levels. China, Brazil, and Mexico are also among the countries highly vulnerable to rising environmental damage and health problems from hazardous waste. Groundwater could be seriously polluted over many years from slow-motion seepage of toxic e-waste. An exercise coordinated by the International Network for Environmental Compliance and Enforcement and the Seaport Environmental Security Network revealed that 54% of the 72 total targeted inspections showed infringements, mostly related to e-waste. The European Parliament has adopted amendments for strengthening the WEEE Directive on waste electrical and electronic equipment, further encouraging recovering and recycling.

In view of increased threats of conflicts triggered by environmental factors, enforcement of international multilateral agreements should be strengthened. Figure 39 reveals significant efforts on ratifications, but more is needed in the area of implementation of the regulations, as well as in developing a global environmental consciousness.

Figure 39. Number of parties to selected multilateral environmental agreements, 1975–2011

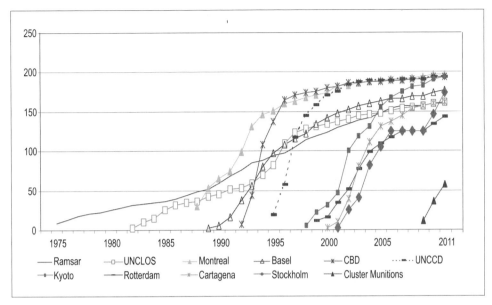

Source: MEA Web sites, with compilation by The Millennium Project

Environmental security analysis should include the impacts of new kinds of weapons; asymmetrical conflicts and warfare; increasing demands on natural resources; urbanization (which makes more people dependent on vulnerable public utilities); environmental degradation and climate change; continued advances in environmental law, with escalating environmental litigations; and the globalization that is increasing interdependencies.

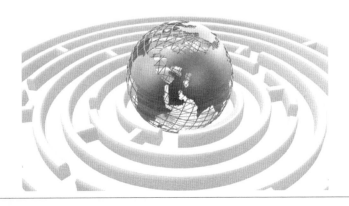

courtesy of: Albertico Acosta http://www.albertico.com/

7

Some Conclusions

The global challenges facing humanity are transnational in nature and transinstitutional in solution. No government, international organization, or other form of institution acting alone can solve the problems described in this report: climate change, cybersecurity threats, organized crime, rich-poor gaps, environmental pollution, international finance, gender discrimination, changing disease situations, and the need for sustainable development. The world may have to move from governance by a mosaic of sometimes conflicting national government policies to a world increasingly governed by coordinated and mutually supporting global policies implemented at national and local levels.

The global financial crisis and the efforts to resolve it clearly demonstrated the need for global systems of analysis, policy formation, and policy implementation. Nation-state decisionmaking has worked well to improve the human condition during slower and less interdependent times. But the future is expected to be far more interdependent—with even less time between problem recognition and solution. We already see that some of the world's toughest problems affect everyone almost instantly and do not end at national borders. In the future, solutions as well as problems must also cross boundaries—ideological as well as geographic. But the future is expected to be far more intensely interdependent, with accelerating changes, than today; hence it will require improved global governance.

Although many people criticize globalization's potential cultural impacts, it is increasingly clear that cultural change is necessary to address global challenges. The development of genuine democracy requires cultural change, preventing the transmission of AIDS requires cultural change, sustainable development requires cultural change, ending discrimination against women requires cultural change, and ending ethnic violence requires cultural change.

The role of the state is more important in countries where there is little private-sector activity and high illiteracy; policies that make sense in more-industrial countries that include leadership from the private sector are less applicable in some poorer regions. Nevertheless, the networks of civil society organizations have to expand to improve governance and decisionmaking at global, regional, and local levels.

There are many answers to many problems, but there is so much extraneous information that it is difficult to identify and concentrate on what is truly relevant. Since healthy democracies need relevant information, and since democracy is becoming more global, the public will need globally relevant information to sustain this trend. Unfortunately, the pursuit of truth on such global matters is not as reinforced as the capacity to rationalize political positions. "Telling truth to power" will be increasingly critical to the future of an ever more sophisticated civilization.

Many leaders in politics, business, and academia have not shown sufficient imagination, courage, or foresight in their proposed solutions to global challenges. Day after day, year after year, decisions that should have been made were not, leading some to believe that we are letting the future fall through our fingers.

Many see the world as a fixed-pie, zero-sum game, with someone's gain becoming another's loss. Others see an expanding pie, grown by new efficiencies and innovations, "a rising tide lifting all boats." And a few others see the world as an exponential growth of pies—with the Internet redistributing the means of production in the knowledge economy, cutting through old hierarchical controls in politics, economics, and finance. They expect a world of unlimited possibilities and think that synergetic analysis will create a better world than decisions based solely on competitive analysis. Countering the "me-first, short-term profits" mind-set may be essential to engaging the world in more-serious consideration of long-term strategies.

Economic growth and technological innovation have led to better health and living conditions than ever before in history for more than half the people in the world, but unless our financial, economic, environmental, and social behaviors are improved along with our industrial technologies, the long-term future is in jeopardy. Business-as-usual is an environmental security threat. Environmental damages that people and organizations got away with in the past are less likely to escape exposure and punishment in the future.

Governments should create systems of resilience and collective intelligence and should use national state of the future indexes for their budget and policy processes. Potential decisionmakers should be trained in decisionmaking and foresight. It is dangerous to have leaders untrained in decisionmaking make irreversible decisions that can affect everyone deeply and almost instantly. Legislatures and parliaments should establish standing permanent committees for the future, as Finland has done.

Creating global partnerships between the rich and poor to make the world work for all seemed like an idealistic slogan before September 11, 2001; now it may prove to be the most pragmatic direction. The world needs a long-term strategic plan for improving the human condition for all.

We cannot ignore the possibility that one day an individual acting alone could create and use a weapon of mass destruction or that there will be serious pandemics as more people and animals crowd into cities while easy transborder travel exists and biodiversity is diminishing. The idealism of the welfare of one being the welfare of all could become a pragmatic long-range approach to countering terrorism, keeping airports open, and preventing destructive mass migrations and other potential threats to human security.

Ridiculing idealism is shortsighted, but idealism untested by the rigors of pessimism can be misleading. The world needs hardheaded idealists who can look into the worst and best of humanity to create and implement strategies of success.

The Millennium Project Nodes

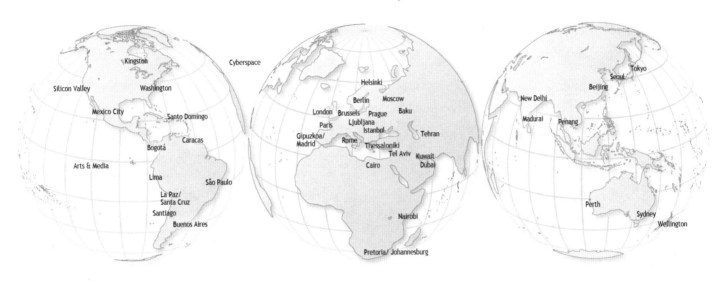

APPENDIX

Millennium Project Participants Demographics

There were 748 futurists, scholars, business planners, scientists, and decisionmakers who contributed this year to the global challenges, State of the Future Index, Latin America 2030, Egypt 2020, and the future of arts and media studies. The following graphs show the regional and sectoral demographics.

Figure 40. Participants in the 2010–11 program

Total participants: 748

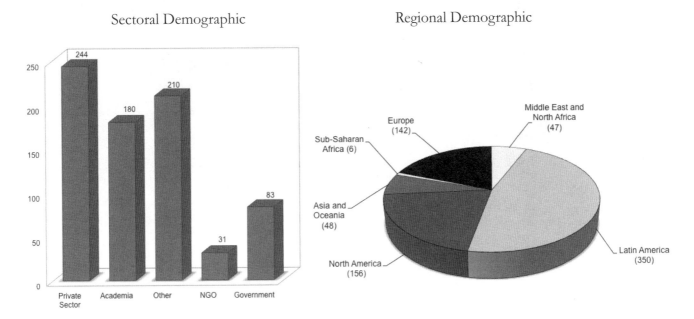

However, much of the work is cumulative in nature, which has come from 4,350 participants over the past 15 years. The second set of graphs shows their regional and sectoral demographics.

Figure 41. Participants since 1996

Total participants: 4,350

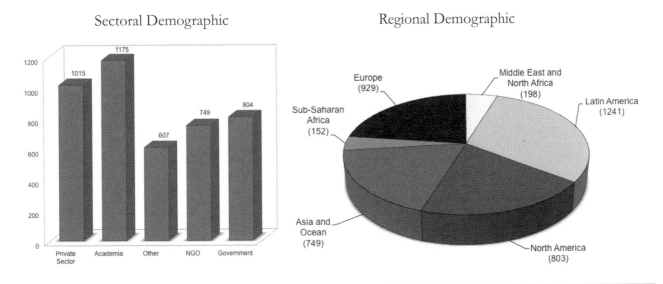

CURRENT AND PREVIOUS SPONSORS

Alan F. Kay & Hazel Henderson Foundation for Social Innovation, St. Augustine, FL (1996–2000)

Amana Institute, São Paulo, Brazil (2004)

Applied Materials, Santa Clara, California (2002–09)

Army Environmental Policy Institute, Arlington, Virginia (1996–2011)

Azerbaijan State Economic University (2009–2011)

City of Gimcheon (via UN Future Forum, South Korea) (2009–10)

Deloitte & Touche LLP, Cleveland, Ohio (1998–09)

The Diwan of His Highness the Amir of Kuwait (2010–11)

Ford Motor Company, Dearborn, Michigan (1996–97, 2005–06)

Foundation for the Future, Bellevue, Washington (1997–98, 1999–2000, 2007–08)

General Motors, Warren, Michigan (1998–2003)

Government of the Republic of Korea (via UN Future Forum) (2007–08)

The Hershey Company (2008–09)

Hughes Space and Communications, Los Angeles, California (1997–98, 2000)

Kuwait Oil Company (via Dar Almashora for Consulting) (2003–04)

Kuwait Petroleum Corporation (via Dar Almashora for Consulting) (2005–06)

Ministry of Communications, Republic of Azerbaijan (2007–11)

Ministry of Education and Presidential Commission on Education, Republic of Korea (2007)

Monsanto Company, St. Louis, Missouri (1996–98)

Motorola Corporation, Schaumburg, Illinois (1997)

Pioneer Hi-Bred International, West Des Moines, Iowa (1997)

Rockefeller Foundation (2008–11)

Shell International (Royal Dutch Shell Petroleum Company), London, United Kingdom (1997)

UNESCO, Paris, France (1995, 2008–10)

United Nations Development Programme, New York, (1993–94)

United Nations University, Tokyo, Japan (1992–95, 1999–2000)

U.S. Department of Energy, Washington, D.C. (2000–03)

U.S. Environmental Protection Agency, Washington, D.C. (1992–93, 1996–97)

Universiti Sains Malaysia (2011)

Woodrow Wilson International Center for Scholars (Foresight and Governance Project), Washington, D.C. (2002)

World Bank (via World Perspectives, Inc.) (2008)

ACRONYMS AND ABBREVIATIONS

AQAL	All Quadrant All Levels
AU	African Union
BRIC	Brazil, Russia, India, and China
CBD	Convention on Biological Diversity
CDC	Centers for Disease Control and Prevention
CITES	Convention on International Trade in Endangered Species of Wild Fauna and Flora
CMS	Convention on Migratory Species
CO_2	carbon dioxide
DAC	Development Assistance Committee (of OECD)
DIY	do-it-yourself
DOTS	Directly Observed Treatment Short course (on TB)
DRC	Democratic Republic of the Congo
DSS	decision support software
EC	European Commission
ECLAC	Economic Commission for Latin America and the Caribbean
EPIC	Established Politics for Information and Communication
EPTA	European Parliamentary Technology Assessment
ETS	Emission Trading Scheme
EU	European Union
FAO	Food and Agriculture Organization of the UN
FDI	foreign direct investment
GDP	gross domestic product
GHG	greenhouse gas
GMO	genetically modified organism
GNI	gross national income
HDI	Human Development Index
HMD	head-mounted display
IAEA	International Atomic Energy Agency
ICC	International Criminal Court
ICT	information and communication technology
IDP	internally displaced persons
IEA	International Energy Agency
IFC	International Finance Corporation
IFPRI	International Food Policy Research Institute
IFs	International Futures
ILO	International Labour Organization
IMF	International Monetary Fund
IMO	International Maritime Organization
IPCC	Intergovernmental Panel on Climate Change
ISO	International Organization for Standardization
IT	information technology
ITPGRFA	International Treaty on Plant Genetic Resources for Food and Agriculture
IWMF	International Women's Media Foundation
LAC	Latin America and the Caribbean

LAU	Latin America University
MDB	multilateral development bank
MDG	Millennium Development Goal
MEA	multilateral environmental agreement
MENA	Middle East and North Africa
MWI	many worlds interpretation
NBIC	nanotechnology, biotechnology, information technology, and cognitive science
NGO	nongovernmental organization
ODA	official development assistance
OECD	Organisation for Economic Co-operation and Development
OSCE	Organization for Security and Co-operation in Europe
OTEC	ocean thermal energy conversion
PEPFAR	President's Emergency Plan for AIDS Relief
ppm	parts per million
PPP	purchasing power parity
RTD	Real-Time Delphi
S&T	science and technology
SCAF	Supreme Council of the Armed Forces (Egypt)
SIMAD	Single Individual Massively Destructive
SOFI	State of the Future Index
TB	tuberculosis
TIA	trend impact analysis
TOC	transnational organized crime
UK	United Kingdom
UN	United Nations
UNCCD	United Nations Convention to Combat Desertification
UNCLOS	United Nations Convention on the Law of the Sea
UNDP	United Nations Development Programme
UNEP	United Nations Environment Programme
UNFCCC	United Nations Framework Convention on Climate Change
UNFPA	United Nations Population Fund
UNIDO	United Nations Industrial Development Organization
UNODC	UN Office on Drugs and Crime
UREDAP	Urgent Educational Attention Program
U.S.	United States
USAID	United States Agency for International Development
WFP	World Food Programme
WHC	World Heritage Convention
WHO	World Health Organization
WMD	weapons of mass destruction
WMO	World Meteorological Organization
WSAI	Wiki Semantic Artificial Intelligence

LIST OF FIGURES, BOXES, AND TABLES

Figures

Boxes

Tables

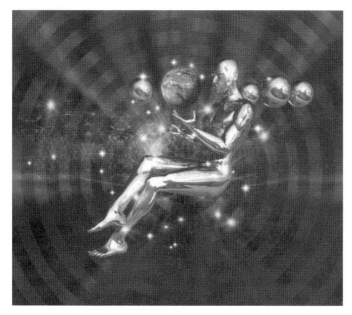

2 0 1 1
FUTURES
World Foresight Encyclopedia Dictionary

Directed and produced by
Concepción Olavarrieta

With assistance from
Jerome C. Glenn and Theodore J. Gordon

Nodo Mexicano. El Proyecto del Milenio and
The Millennium Project 2011

ISBN: 978-0-9818941-6-4
Format: print and CD
Price: $ 49.99

Welcome to the World of Foresight! *FUTURES* invites you to discover the fascinating world of futurists. Through each term or method you will become more familiar with the specific jargon of futures studies and will learn how to build desirable, possible, or catastrophic futures.

Over the past several years, more than 3,000 futurists from The Millennium Project network contributed terms and methods used by futures studies, which got organized in this compendium in a friendly way, so that they could become part of your daily vocabulary for sharing dreams and views with other futurists from around the world.

This comprehensive bilingual dictionary/encyclopedia contains over 880 terms of the domain of futures studies.

Each term presents details on the name, etymology, sense, and sources in English and Spanish. The over 650 pages anthology will be available in print and on a CD for more convenient search and use.

For further information: http://www.millennium-project.org/Encyclopedia.html

----

To order send a check payable to **The Millennium Project** for $49.99 per copy (plus $6.00 shipping within the US; $12.00 outside the US) or send your credit card information using the form below or via email.

The Millennium Project
4421 Garrison Street, NW
Washington, DC 20016   USA
Tel./Fax: 1-202-686-5179   Discount 40% for orders of 10 copies or more
Email: mp-orders@millennium-project.org
URL: www.millennium-project.org

Name: …………………………………….. Email: …………………………………….

Shipping Address: ………………………………………………………………………….

City: …………………....... State: …………… Zip/Postal Code: ………………………

Credit Card #: …………………………………………………. Exp. Date: ……………..

Futures Research Methodology Version 3.0 is the largest, most comprehensive collection of internationally peer-reviewed methods and tools to explore future possibilities ever assembled in one resource.

Over half of the chapters were written by the inventor of the method or by a significant contributor to the method's evolution.

The CD-ROM contains 39 chapters totaling about 1,300 pages.

Each method is treated in a separate file in word (.doc) and PDF format.

ISBN: 978-0-9818941-1-9
Price: $49.50 US dollars plus shipping

39 chapters of comprehensive handbook on Futures Research Methodology

For further information: http://www.millennium-project.org/millennium/FRM-V3.html

----

To order send a check payable to **The Millennium Project** for $49.50 per copy (plus $6.00 shipping within the US; $12.00 outside the US) or send your credit card information using the form below or via email.

The Millennium Project
4421 Garrison Street, NW
Washington, DC 20016   USA
Tel./Fax: 1-202-686-5179
Email: mp-orders@millennium-project.org
URL: www.millennium-project.org

Discount 40% for orders of 10 copies or more
25% on Version 3.0 to those who bought 2.0

Name: ……………………………….. Email: ………………………………

Shipping Address: …………………………………………………………………

City: ………………........ State: …………… Zip/Postal Code: ………………………

Credit Card #: ……………………………………………. Exp. Date: ……………..

NOTES